Smoothed Particle Hydrodynamics

Carlos Alberto Dutra Fraga Filho

Smoothed Particle Hydrodynamics

Fundamentals and Basic Applications in Continuum Mechanics

Springer

Carlos Alberto Dutra Fraga Filho
Federal Institute of Education, Science and
Technology of Espírito Santo
Vitória, Espírito Santo, Brazil

ISBN 978-3-030-00772-0 ISBN 978-3-030-00773-7 (eBook)
https://doi.org/10.1007/978-3-030-00773-7

Library of Congress Control Number: 2018955176

This Springer imprint is published by the registered company Springer Nature Switzerland AG
The registered company address is: Gewerbestrasse 11, 6330 Cham, Switzerland

[...] Not to the strong is the battle,
Not to the swift is the race,
Yet to the true and the faithful
Vict'ry is promised through grace.

Victory Through Grace
Fanny Jane Crosby (1820–1915)

This book is dedicated to my family.

Preface

This book is the result of studies dedicated to the Lagrangian particle modelling and especially the Smoothed Particle Hydrodynamics (SPH) method over the last decade.

It all began in 2009 when the understanding of the Lagrangian approach was necessary to begin the development of a purely Lagrangian model of particles for the simulation of oil spreading on a calm sea. The studies have come a long way until this final application could be performed.

Several issues arose when attempts to implement the SPH method were being performed. Through the process of implementation, the understanding of the differential equations that govern fluid flow and energy transport were necessary. In this context, the continuum mechanics and its foundations were the object of investigation and support for the search of solutions in the macroscopic scale, and that they did not violate the laws of classical Physics.

Physical-mathematical conclusions reached during the process of study and implementation of the method are brought to the fore. Particular attention is given to the presentation of results and conclusions regarding the boundary conditions, which constitute one of the main challenges of the Lagrangian continuum method.

The objective of this work is to help the researcher/student in the process of the implementation of SPH method for solving problems (in engineering, physics, applied mathematics, biology, geoscience, environmental and other areas of science).

The understanding and modelling of many phenomena require the development of studies in a challenging interface area, where occurs the intersection of multiple disciplines, such as physics, mathematics, chemistry, material sciences. In problems defined in the continuum scale, the classical laws of physics must be applied. On the other hand, in the molecular scale, the quantum physics and the molecular dynamics are applied. There is a gap between these two scales, and a scientific effort has been made aiming to couple the events in the continuum and molecular domains. Hybrid/multiscale models have been developed with the objective of providing solutions to problems in the mesoscopic region (between macroscale and microscale). Within this context, there is a need for a theoretical discussion about the appropriate modelling to be applied to each problem studied.

Considering the physical-mathematical approach of this work, but recognizing the need to integrate it with the computing, the algorithm and computational routines used for the solution of differential equations of conservation discretized by the SPH are presented in flowcharts in the chapter dedicated to the solution of some physical/engineering problems. In the final pages of the book, the computational FORTRAN code used in the solution of the first problem proposed in this work is made available to the reader for use. Additionally, the reader is provided with an open access literature reference, in which the description of the implemented computational code can be found.

This book is intended for academic researches, graduate and postgraduate students in the fields of engineering, physics/continuum mechanics, mathematics/applied mathematics, chemistry, environmental sciences, geosciences, oceanography and computational methods. As prerequisites for the understanding of the contents presented, it is recommended that the reader has prior knowledge of differential and integral calculus, classical physical laws/continuum mechanics, computational fluid mechanics and transport phenomena. This being a multidisciplinary work, knowledge of basic concepts of molecular dynamics is also important for an understanding of the topics covered. However, in order to allow the reader to have access to complementary information on the topics addressed, literature references are provided throughout the text.

In the whole work, care was taken to highlight the most important points and facilitate the understanding of the contents by the reader. At the beginning of each chapter, there is an abstract where the topics that will be detailed on the following pages are listed. Boxes were placed around important equations and results. Concluding remarks/comments are presented at the end of the chapters. An auxiliary material is provided in the appendices, in order to allow an easier understanding of the solutions of the proposed problems.

Without major pretensions, I expect this book to be a support for the study and a simple and current reference on the application of Smoothed Particle Hydrodynamics in continuum fluid mechanics and transport phenomena. And finally, I hope it is a motivation for the advancement of research and instigate the search for satisfactory answers to the challenges currently presented in the application of the SPH method in continuum domain.

Vitória, Brazil
Summer 2018

Carlos Alberto Dutra Fraga Filho

Acknowledgements

I would like to thank Profs. Julio Tomás Aquije Chacaltana and Maxsuel Marcos Rocha Pereira (Federal University of Espírito Santo, Brazil) who followed the stages of understanding the SPH method and the questions which motivated the preparation of this work.

I express my sincere thanks to Prof. Peter Wriggers, Chairman of the V International Conference on Particle-Based Methods (PARTICLES 2017), which occurred at Leibniz Universität Hannover (Germany), and Editor in Chief of Computational Particle Methods, a journal to which I express my gratitude for giving researchers of particle methods the opportunity to present their results and scientific contributions.

I also wish to express my sincere thanks to Prof. Benedict Rogers (University of Manchester, UK), Chair of the Steering Committee of SPH European Research Interest Community (SPHERIC), for providing space for the dissemination of scientific works and opportunities for young scientists on the group's website, through which researchers from around the world can present their results and be in touch with the latest advances in researches and computational tools.

Finally, I want to thank my daughter Amanda and all the other people involved in revisions performed in the text of this work.

Carlos Alberto Dutra Fraga Filho

Contents

1 Introduction 1
- 1.1 General 1
- 1.2 Scientific Method of Problem-Solving 3
- 1.3 Objective 5
- 1.4 Presentation of the Remaining Chapters 5
- References 6

2 Physical-Mathematical Modelling 11
- 2.1 The Continuum Hyphothesis 11
- 2.2 Physical Laws of Conservation 12
- 2.3 Pressure Modelling 13
  - 2.3.1 Equation of State for Dynamic Pressure 13
  - 2.3.2 Modified Pressure 14
- 2.4 Specific Internal Energy Modelling 15
- References 15

3 Smoothed Particle Hydrodynamics Method 17
- 3.1 Fundamentals 17
- 3.2 Discretization of the Continuum Domain 19
  - 3.2.1 Approximation of the Divergent of a Vectorial Function 20
  - 3.2.2 Approximation of the Gradient of a Scalar Function 22
  - 3.2.3 Approximation of the Laplacian 24
  - 3.2.4 SPH Approximations to the Conservation Equations 25
  - 3.2.5 Errors in SPH Approximations 26
  - 3.2.6 Smoothing Functions 26
  - 3.2.7 Neighbouring Particles Search 28
  - 3.2.8 Treatment of the Free Surface 29
  - 3.2.9 Treatment of the Interfaces 35
  - 3.2.10 Turbulence 37
  - 3.2.11 Variable Smoothing Length 39
  - 3.2.12 Numerical Aspects and Corrections 40

3.3 Temporal Integration Methods 43
3.3.1 Euler's Integration Method 44
3.3.2 Leapfrog Method 45
3.3.3 Predictor–Corrector 45
3.4 Consistency 46
3.4.1 Restoration of the Consistency 48
3.5 Boundary Treatment Techniques 53
3.5.1 Fictitious Particles and Artificial Repulsive Forces 54
3.5.2 Dynamic Boundary Conditions 56
3.5.3 Reflective Boundary Conditions 56
3.5.4 Open Periodic Boundary Conditions 60
3.5.5 Additional Remarks 61
References 62

4 Applications in Continuum Fluid Mechanics and Transport Phenomena 67
4.1 Procedure Employed in Problem-Solving 67
4.2 Heat Diffusion in a Homogeneous Flat Plate 68
4.2.1 Physical-Mathematical Modelling 68
4.2.2 Numerical Simulations 71
4.2.3 Comments 76
4.3 Still Liquid Inside an Immobile Reservoir 78
4.3.1 Physical-Mathematical Modelling 78
4.3.2 Numerical Simulations 80
4.3.3 Comments 83
4.4 Dam Breaking Over a Dry Bed 84
4.4.1 Physical-Mathematical Modelling 84
4.4.2 Numerical Simulations 85
4.4.3 Comments 88
4.5 Oil Spreading on a Calm Sea 89
4.5.1 Motivation 89
4.5.2 Physical-Mathematical Modelling 91
4.5.3 Numerical Simulations 92
4.5.4 Results and Discussions 96
4.6 Concluding Remarks 98
References 99

Computer Code 101

Conclusion 105

Appendix A: Smoothing Functions, Derivatives and Normalization Constants 107

Appendix B: Deduction of the SPH Laplacian Operator 115

FORTRAN Source Files 121

References 143

Index 145

About the Author

Carlos Alberto Dutra Fraga Filho was born in Vitória, Espírito Santo, Brazil. He holds the following degrees: undergraduate degree in Mechanical Engineering (1998), Master of Science in Mechanical Engineering (2007) and Doctor of Science in Environmental Engineering (2014) at Federal University of Espírito Santo, Brazil. He has experience in mechanical and environmental engineering, with emphasis on computational fluid dynamics, working mainly in the following areas: Lagrangian modelling and Smoothed Particle Hydrodynamics (SPH) method, fluid mechanics and transport phenomena. He is Professor of Mechanical Engineering, Journals' Reviewer—*Physics of Fluids* (American Institute of Physics), *Journal of Mechanics Engineering and Automation* and *Engineering Teaching Journal* (Brazilian Association of Engineering Teaching)—and leader of the research group "Development, Implementation and Application of Computational Tools for the Solution of Problems in Engineering", certified by the Coordination for the Improvement of Higher Education Personnel (CAPES).

Symbols

The next list describes the symbols that will be used in this book.

Roman Symbols

$\tilde{A}$	matrix used in the MLS correction method
A_N	coefficients in the series
B	term associated with fluctuations in the fluid density in Tait equation
C^0	consistency of zeroth order
C^1	consistency of first order
$(\mathrm{C}_1)_N$	coordinate of the mass centre position, perpendicular to the plane collision, at the time instant $(to + \Delta t)$, in a movement without obstacles, before the reflection
$(\mathrm{C}_\mathrm{f})_N$	coordinate of the mass centre position, perpendicular to the plane collision, at the time instant $(to + \Delta t)$, in a movement without obstacles, after the reflection
c	magnitude of sound velocity propagation in the fluid
c_i	velocity of sound in the fixed particle
c_j	velocity of sound in the neighbouring particle
c_{ij}	average of the sound velocities in the fixed and neighbouring particles
CF	coefficient of friction
CR	coefficient of restitution of kinetic energy
Cs_i	value of the colour field at the fixed particle
Cs_j	value of the colour field at the neighbouring particle
c_v	specific heat at constant volume
D	oil slick diameter
d	distance between the particle's mass centre and the collision plane
$d\mathbf{S}$	infinitesimal surface element
dX'	infinitesimal volume element

dx	horizontal distance between two adjacent centres of mass at the initial particle disposition
dy	vertical distance between two adjacent centres of mass at the initial particle disposition
$\frac{D}{Dt}$	Lagrangian (or material) derivative
$\frac{De_i}{Dt}$	rate of change of specific internal energy of the fixed particle
$\frac{De_i{}^*}{Dt}$	rate of change of specific internal energy of the fixed particle in time after the addition of the term related to the artificial viscosity to the energy conservation equation
$\frac{D\vec{v}_i}{Dt}$	acceleration of the fixed particle
$\frac{D\vec{v}_i{}^*}{Dt}$	acceleration of the fixed particle, after the addition of the term related to the artificial viscosity to the momentum conservation equation
e	specific internal energy
e_i	specific internal energy of the fixed particle
$e_i(t_o + \Delta t)$	specific internal energy of the fixed particle at the final time instant
$\mathbf{f}_{ab}$	restorative contact force
$\vec{f}_{ext}$	external forces acting on the fluid, per unit volume
f_i	approximate value of the function in the fixed particle's position
f_j	approximate value of the function in the neighbouring particle position
$\mathbf{f}_i$	vectorial function evaluated at the position occupied by the fixed particle
$\mathbf{f}_j$	vectorial function evaluated at the position occupied by the neighbouring particle
$f(X)$	function value at the fixed point X
$f(X')$	function value at the variable point X'
$f'(X)$	first-order derivative of the function evaluated in the position X
$f''(X)$	second-order derivative of the function evaluated in the position X
$\mathbf{f}(X)$	vectorial function evaluated at the position of the fixed point
$\mathbf{f}(X')$	vectorial function evaluated at the position of the variable point
$< f(X) >$	approximated value of the scalar function f at the fixed point X
$\mathbf{fs}(X_s)$	force of surface tension acting in the direction normal to the free surface, at the point of surface X_s
$\mathbf{fs}_i$	force of surface tension exerted in the position of the particle surface i, in the direction normal to the interface
$\vec{g}$	gravity
h	smoothing length
H	ordinate of the fluid surface above the point analysed
h_i	smoothing length of the fixed particle

h_i	smallest smoothing length (employed in the choice of the time step)
h_j	smoothing length of the neighbouring particle
h_{ij}	average of the smoothing length (fixed and neighbouring particle)
h_0	initial smoothing length
h_o	initial oil height
i	fixed particle
j	neighbouring particle
K	thermal conductivity
k	scaling factor that depends on the kernel used
kh	support radius (radius of the circumference of the domain of influence)
k_1	constant in the Fay's equation
L_i	inverse of the matrix M_i in the kernel gradient correction method
l	threshold parameter
M_i	matrix used in the kernel gradient correction method
M_N	kernel moment of order N
m	current time step
m'	mass of the infinitesimal volume element of the fluid
m_j	mass of the neighbouring particle
Ma	Mach number
N	number of points employed in the series
n	number of neighbouring particles
n_d	number of domain dimensions
n_p	number of particles arranged on each side of the flat plate
N_1	parameter used in the calculation of the restorative contact force
N_2	parameter used in the calculation of the restorative contact force
n_1	parameter used in the calculation of the Lennard-Jones force
n_2	parameter used in the calculation of the Lennard-Jones force
$\mathbf{n}$	unit vector normal to the surface
$n(X_s)$	vector normal to the free surface located at the point X_s, in the interface
n_i	vector normal to the free surface, located in the position of the particle surface i
$(n_u)_i$	unit vector normal to the free surface, located in the position of the particle surface i
P	absolute pressure
$\bar{P}$	coefficient to the calculation of the restorative contact force
P_i	absolute pressure acting on the fixed particle
P_j	absolute pressure acting on the neighbouring particle

P_{dyn}	dynamic pressure
P_{mod}	modified pressure
$\mathrm{P}_{\mathrm{mod}_i}$	modified pressure on the fixed particle
$\mathrm{P}_{\mathrm{mod}_j}$	modified pressure on the neighbouring particle
P_o	reference pressure
$P_i\left(t_o + \frac{\Delta t}{2}\right)$	pressure on the fixed particle calculated in the middle of the integration time step
$P(\nabla.\vec{\mathrm{v}})$	work rate due to pressure forces, per unit volume
q	ratio of the distance between a fixed and a variable point to the smoothing length
$\dot{q}$	heat rate provided by the source, per unit volume
R	parameter used in the detection of the particle–particle contact
R_p	variable used in the calculation of the artificial pressure
$\mathrm{Re}(h^2)$	residue of second order, after the truncation of the Taylor series
r	radial direction
r_{ab}	distance between the centres of mass of two fluid particles at the interface
$\mathbf{r}_i$	position of the fixed particle
$\mathbf{r}_j$	position of a neighbouring particle
$\mathbf{r}_{ij}$	radial distance between the fixed and a neighbouring particle
$\lvert r_{iv} \rvert$	distance between the centres of mass of a fluid and a virtual particle in the calculation of the Lennard-Jones force
$\mathrm{r_p}$	particle radius
r_0	cut-off distance
S	integration surface
s	parameter used in the calculation of the artificial pressure
T	absolute temperature
T_i	temperature of the fixed particle
T_j	temperature of the neighbouring particle
T^m	particle's temperature in the current time step
T^{m+1}	prediction of the particle's temperature obtained from the numerical integration method
T_S	temperature at the lower boundary
T_{series}	temperature obtained by series
T_{SPH}	temperature provided by SPH solution
$T(\mathrm{x}, y)$	temperature distribution in the flat plane
t	time
t_o	initial instant of time
t_f	final time instant of the gravity-inertial spreading
V	volume of the oil spilled
v	magnitude of the flow velocity
$\vec{\mathrm{v}}$	velocity
$(\mathrm{v_{col}})_N$	magnitude of the component of the particle velocity perpendicular to the collision plane, before the collision

$(\mathrm{v}_{\mathrm{col}})_T$	magnitude of the component of the particle velocity tangential to the collision plane, before the collision
$\vec{\mathrm{v}}_i$	velocity of the fixed particle
$\vec{\mathrm{v}}_i^*$	velocity of the fixed particle, after the application of the XSPH correction
$\vec{\mathrm{v}}_j$	velocity of the neighbouring particle
$\vec{\mathrm{v}}_{ij}$	relative velocity between the fixed and neighbouring particles
$\vec{\mathrm{v}}_i^P$	prediction of the velocity of the fixed particle
$\vec{\mathrm{v}}_i(t_o + \Delta t)$	velocity of the fixed particle at the final time instant
$(\mathrm{v}_\mathrm{p})_T$	magnitude of the component of the particle velocity tangential to the collision plane, immediately after the collision
$(\mathrm{v}_\mathrm{p})_N$	magnitude of the component of the particle velocity perpendicular to the collision plane, immediately after the collision
X	position of the fixed point
X'	position of the variable point
X_{ab}	difference between the position vectors of the particles of fluids A and B
X_i	fixed particle's position
X_j	neighbouring particle's position
x_i	abscissa of the fixed particle
x_j	abscissa of the neighbouring particle
y_i	ordinate of the fixed particle
y_j	ordinate of the neighbouring particle
X_{ij}	difference between the position vectors of the fixed and neighbouring particles
X_i^P	prediction of the position of the fixed particle
X_{iv}	difference between the position vectors of the virtual particle and the fluid particle
$X_i(t_o + \Delta t)$	position of the fixed particle at the final time instant
$\mathrm{x}_{\mathrm{exp}}$	abscissa of the wave front obtained from experiments
x_{SPH}	abscissa of the wave front provided by SPH simulations
$W(r, h)$	smoothing function written in polar coordinate system
$W(\mathbf{r}_i - \mathbf{r}_j, h)$	smoothing function written in polar coordinates, evaluated at the position $(\mathbf{r}_i - \mathbf{r}_j)$
$W(X - X', h)$	kernel or smoothing function evaluated at the position $(X - X')$
$W(X_i - X_j, h)$	kernel evaluated at the position $(X_i - X_j)$
$W'(X' - X, h)$	derivative of the kernel evaluated in position $(X' - X)$
$W^{MLS}(X_i - X_j, h)$	smoothing function provided by the mean least square (MLS) correction method evaluated at the position $(X_i - X_j, h)$

Greek Symbols

α	thermal diffusivity
α_i	thermal diffusivity of the fixed particle
α_D	constant that assures the kernel normalization over the domain of influence
α_π	constant used in the calculation of artificial viscosity
β	free surface parameter
β_π	constant used in the calcultaion of artificial viscosity
$\beta(X_i)$	correction vector used in the mean least square (MLS) method
γ	parameter in the Tait equation
ε_{v}	energy dissipation rate per unit volume
η	parameter ranging from 0 to 1.0 used in the XSPH correction
$\Delta\mathrm{D}$	differences between the predicted oil slick diameters (Fay's equation—SPH result)
Δf_i	Laplacian of the function at the position of the fixed particle
Δp	average spacing in the neighbourhood of the fixed particle
ΔT_p	largest percentage difference between temperatures, obtained by series and provided by SPH, in a defined spatial point
Δ_{w}	relation between oil and water densities
ΔT	temperature error
ΔT_p	largest percentage difference between temperatures in a defined spatial point
ΔV_j	volume of each Lagrangian neighbouring particle
$\Delta \mathrm{x_p}$	percentage difference between the abscissas of the wave fronts
$\delta(X - X')$	Dirac delta function evaluated at the position $(X - X')$
$\delta_{l,t}$	Kronecker delta function
κ	curvature of the free surface
λ	parameter used in the calculation of the artificial pressure
μ	absolute viscosity
$\mu\nabla^2\vec{\mathrm{v}}$	viscosity forces per unit volume
π_{ij}	artificial viscosity
ρ	density
ρ_o	fluid density at rest
$\rho\vec{g}$	gravitational force per unit volume
ρ'	density of the infinitesimal volume element of the fluid
ρ_i	density of the fixed particle
$(\rho_i)_{corrected}$	density of the fixed particle after the MLS correction
$(\rho_i)_{filtered}$	density of the fixed particle after the density renormalization
ρ_{j}	density of the neighbouring particle
$\bar{\rho}_{ij}$	average of the densities (fixed and neighbouring particle)
ρ_{air}	density of the air
ρ_o	density of the oil
ρ_{w}	density of the water

ρ_{water}	density of the water inside the immobile reservoir
ρ_*	density of the particle in the current numerical iteration
ρ_0	density of the particle in the previous numerical iteration
ρ_{zero}	initial density, used to update the smoothing length
ρ_i^P	prediction of the density of the fixed particle
$\rho_i(t_o + \Delta t)$	density of the fixed particle at the final time instant
$\left(\frac{1}{\rho}\frac{\partial P}{\partial \mathrm{x}}\right)_i^*$	component of the pressure gradient in the fixed particle, corrected by the CSPM method, per unit mass, in the direction x
$\left(\frac{1}{\rho}\frac{\partial P}{\partial \mathrm{y}}\right)_i^*$	component of the pressure gradient in the fixed particle, corrected by the CSPM method, per unit mass, in the direction y
σ	fluid surface tension coefficient
υ_{air}	kinematic viscosity of the air
υ_i	kinematic viscosity of the fixed particle
υ_{water}	kinematic viscosity of the water inside the reservoir
Ψ	parameter that depends on each problem and must be of the same order of magnitude as the square of the highest flow velocity
φ^2	factor that prevents numerical differences when two particles approach one another
χ_{ij}^2	variable used in the calculation of the artificial viscosity
Ω	domain of influence
∇	mathematical nabla operator
∇Cs_i	gradient of the smoothed colour field at the position of the surface particle i
∇f_i	gradient of the function evaluated in the position occupied by the fixed particle
$<\nabla f(X)>$	approximation to the gradient of the function at the position X
$\nabla.\mathbf{f}_i$	divergent of the vectorial function in the fixed particle's position
$\nabla.\mathbf{f}(X)$	approximation for the divergent at the position X
∇P	pressure forces per unit volume
$\nabla \mathrm{P}_{\mathrm{mod}}$	modified pressure gradient
$\nabla.\mathbf{r}_i$	SPH divergence of the fixed particle's position
∇_s	gradient of the surface
$\nabla \cdot (K\nabla T)$	heat rate due to the diffusion per unit volume
$\nabla W_{ij}^{\mathrm{x}}$	component of the kernel gradient in the direction x
$\nabla W_{ij}^{\mathrm{y}}$	component of the kernel gradient in the direction y
$\nabla^2 f_i$	is the SPH approximation of the Laplacian of the function f evaluated at the position occupied by the fixed particle
$\tilde{\nabla} W_{ij}$	gradient provided by the kernel gradient correction method
$\tilde{\nabla} W_{ij}^{\mathrm{x}}$	component of the corrected kernel gradient in the direction x
$\tilde{\nabla} W_{ij}^{\mathrm{y}}$	component of the corrected kernel gradient in the direction y

List of Figures

Fig. 1.1	Steps in the scientific method of problems solving	4
Fig. 3.1	Graphical representation of the kernel and its domain of influence. The reference particle 'i' (black central circle) has its physical properties influenced by the neighbour particles 'j', which are within the domain of influence	18
Fig. 3.2	A two-dimensional cell with 21 particles within the domain of influence	21
Fig. 3.3	The search for neighbouring particles: **a** directly and **b** using a grid	29
Fig. 3.4	Flow chart showing the operations performed in a linked list code	30
Fig. 3.5	The cut-off radius used in the Verlet list ($kh + L$) and the probable neighbour particles storaged in the neighbour list	31
Fig. 3.6	Directions of force of surface tension at points with varying curvature	31
Fig. 3.7	Illustration of the material interface. **a** Initial distribution and **b** evolution of the particles	36
Fig. 3.8	Illustration of the contact and penetration between particles	36
Fig. 3.9	Domain of influence in an two-dimensional case: **a** complete and **b** truncated	47
Fig. 3.10	Schematic illustration of the solid contour region. Arrangement of the virtual particles of type I (a line on the contour) and of the virtual particles of type II (in an extended area beyond the domain)	54
Fig. 3.11	Interaction between the dynamic particles fixed on the contour and a fluid particle that approaches the border (full black circle)	56
Fig. 3.12	The velocities of particles before the impact against the plane (A) and immediately after it (B)	59
Fig. 3.13	Collisions experienced by a particle in a time step	59

Fig. 3.14 Periodic open boundary conditions. The domain of influence of the particle 'i' is extended to the complimentary open lateral boundary 61
Fig. 4.1 Procedure employed to obtaining a numerical solution to a physical/engineering problem 68
Fig. 4.2 Homogeneous flat plate with initial and boundary conditions 70
Fig. 4.3 Algorithm implemented for diffusion in a homogeneous flat plate 71
Fig. 4.4 Initial setup of 50 × 50 particles distributed inside the domain and 100 particles on each side of the plate 72
Fig. 4.5 Heat diffusion at different instants of time, until the steady state is reached (spline cubic kernel, 2,500 particles) 73
Fig. 4.6 Effect of increasing the particle number in the domain. Solution by using the quintic spline kernel: **a** 2,500, **b** 4,900 and **c** 8,100 particles 77
Fig. 4.7 Behaviour of the highest relative error 78
Fig. 4.8 Hydrostatic pressure field 79
Fig. 4.9 Flowchart of the numerical code for the reservoir problem, employing the reflective boundary conditions 81
Fig. 4.10 Open reservoir containing a Newtonian, uniform, incompressible and isothermal fluid 81
Fig. 4.11 Initial setup of particles and their centres of mass (represented by dots) and the solid boundaries 82
Fig. 4.12 Particles at rest inside the reservoir. The squares and the plus signs represent the centres of mass of the fluid particles at the time instant 0 and at the end of the simulation, respectively 83
Fig. 4.13 Flowchart of the algorithm for dam breaking simulation 85
Fig. 4.14 Simulated 2-D computational domain and the initial distribution of the centres of mass of the particles 86
Fig. 4.15 Schematic diagram of the experimental apparatus 87
Fig. 4.16 Comparison between the experimental and numerical results 88
Fig. 4.17 Main physical, chemical and biological processes occurring at an oil slick in the sea 90
Fig. 4.18 Oil and the water particles separated by a rough and rigid plane 91
Fig. 4.19 Schematic 3-D illustration of the oil cylinder over the water surface at the initial instant of the spill 92
Fig. 4.20 Longitudinal section of the oil cylinder at the initial time instant 93
Fig. 4.21 Upper view: the circular shape of the oil slick spreading on a calm sea condition 93

Fig. 4.22 Domain of influence of an oil particle near the interface. The water particles neighbouring the reference oil particle were identified (All particles are represented by their centres of mass) 94
Fig. 4.23 Evolution of the reach of the oil slick in time. The inner circumference refers to the initial diameter of the oil (at the beginning of the spill) 94
Fig. 4.24 Flowchart of the algorithm used in the oil spreading simulations 95
Fig. 4.25 Simulations results for CR 0.85 and 0.90 and CF equal to 0.15 and 0.10 97
Fig. 4.26 Simulations results for CR 0.92 and CF equal to 0.15 and 0.10 98

List of Tables

Table 4.1 Initial configurations of particles and lateral distances of the centres of mass. 71

Table 4.2 Largest percentage difference between temperatures provided by series and SPH solution 74

Table 4.3 Positions of the particles where the greatest differences between temperatures occurred. 75

Table 4.4 Oil slick diameter at the end of the gravity-inertial stage 96

Introduction

1

Abstract

During the studies development and implementation of particle method to the solution of engineering and physical problems, several questions and doubts arise. Although the physical-mathematical analysis should be carried out in the continuum macroscopic domain, the concepts of molecular dynamics (mainly in the application of boundary conditions and interface treatment) and continuum mechanics are confused. This book has been written based on results obtained over a decade of study and research. During this time, in order to achieve good results without making conceptual mistakes, the employment of dynamic molecular models in the continuum scale has been questioned. In this context, the continuum mechanics fundamentals must be the support for the domain discretization and search of solutions using the Smoothed Particle Hydrodynamics (SPH) method, and the classical laws of physics cannot be violated. Considering this need, this work brings consistent solutions to hydrostatic and hydrodynamics problems.

Keywords

Macroscopic scale · Continuum hypothesis · Particle method
Molecular modelling · Scientific method

1.1 General

The Eulerian view is historically the most commonly used in the modelling of physical problems involving fluids and transport phenomena. This modelling requires the use of numerical methods employing grids or meshes, such as finite differences, finite volumes and finite elements, for the solution of partial differential equations (PDEs) or integral formulations [1]. The greatest difficulty in using meshes is ensuring consistency. Employing meshes can lead to a number of difficulties when dealing with fluid flow problems with free surface, mobile interfaces, complex geometries

C. A. Dutra Fraga Filho, *Smoothed Particle Hydrodynamics*,
https://doi.org/10.1007/978-3-030-00773-7_1

and topological changes. If these features are present, generating a mesh at each numerical iteration (remeshing), a prerequisite for the numerical simulation of quality becomes a difficult process, both time-consuming and expensive.

Lagrangian modelling is based on the hypothesis that the problem domain can be divided into a finite number of particles that, in its original conception, do not interact with one another (one-particle model). The centre of mass of each particle receives the coordinates that define its position in space, and the physical properties of each Lagrangian element are found at each instance in time.

The meshless Lagrangian particle model is being increasingly used in research, providing stable and accurate numerical solutions for PDEs or integral equations. Their main advantages over the Eulerian modelling are:

- Free surfaces and their topological changes, interfaces and mobile boundaries can be naturally traced by particles in the simulation process.
- Simplicity and lower computational cost in complex geometries.
- No need for remeshing (one characteristic of mesh methods) at each numerical iteration, which decreases the computational cost in the remeshing operations.
- Conservation of mass (there is no mass flow through the Lagrangian element).
- The history of the particles can be easily obtained, and their graphical visualization of the results is available, allowing a better understanding of the spatio-temporal evolution of the fluid and its properties.
- Control of undesirable numerical diffusion, a characteristic problem of methods based on meshes, due to conservation of mass at the Lagrangian element.

Over the last few years, meshless methods have been an alternative applied to the studies, pointing to a new generation of more effective computational methods for solving more complex problems. A good presentation of the particle methods starting with the Smoothed Particle Hydrodynamics (SPH), followed by the least squares (MLS) method and the Galerkin method, is presented by Huerta and collaborators [2].

Until a few decades ago, Lagrangian particle modelling was not used in the simulation of problems in fluid mechanics and transport phenomena. In 2004, Fox, McDonald and Pritchard [3] stated that the employment of the Lagrangian approach to analyse a fluid flow and keeping track the motion of individual particles of fluid would be impractical, advising the use of another type of modelling in the study of fluids.

For a long time, Lagrangian modelling was put aside in favour of Eulerian methods due mainly to the processing time: in general, higher than that required by mesh methods. Currently, however, the progress of the processing techniques, the rapid spread of multicore processors, and parallelization, made the use of particle methods possible. Graphics processing units (GPUs), using Compute Unified Device Architecture (CUDA), developed by NVIDIA, allow the use of more than one million particles in the discretization of the fluid domain [4].

Most graduate and undergraduate students still study physical laws according to the Eulerian view, basically due to the academic domain of numerical methods that use meshes to solve the differential equations used in problem modelling. However, nowadays, the meshless particle methods have become an alternative for the solution of the physical problems, with a promising future, due mainly to the advance of the processing techniques.

Nowadays, SPH method is applied to diverse areas and problems of the solid and fluid mechanics. In the next lines, some applications with references will be presented:

- Free surface flows [5–8].
- Wave propagation [9–11].
- Floating objects [12–14].
- Wave–structure interactions [15–17].
- Sloshing [18–20].
- Floods and tsunamis [21–24].
- Lava flows [25–27].
- Multiphase flows [28].
- Soil–water interaction/hydraulic erosion [29–32].
- Sediment transport [33–35].
- Flows in porous media [36–39].
- Impacts on solids [40–42].
- Heat diffusion [43–45].
- Fracture [46–48].
- Casting [49–51].
- Industrial applications [52,53].
- Biological tissue [54].
- Medicine and biomechanics [55–61].
- Electromagnetism [62,63].
- automotive industry [52,64–66]
- lubrication and tribology [67–69]
- aeronautics [70–74]
- navigation and swimming [75,76]

Monaghan [77] presented other applications, such as dam breaks and plunging waves, landslides and bodies impacting fluids and granular flow.

The learning of the Lagrangian modelling and particle methods is a current need in the courses of engineering, physics, mathematics, oceanography, environmental sciences and geosciences. In addition, there is an increase in researches that aim at the development of Lagrangian mathematical models and new computational tools for solving problems.

1.2 Scientific Method of Problem-Solving

Brief comments will be made on the scientific method of problem-solving (employed in problems presented in this book).

To obtain the solution of any problem in science and engineering, the researcher should go through the steps:

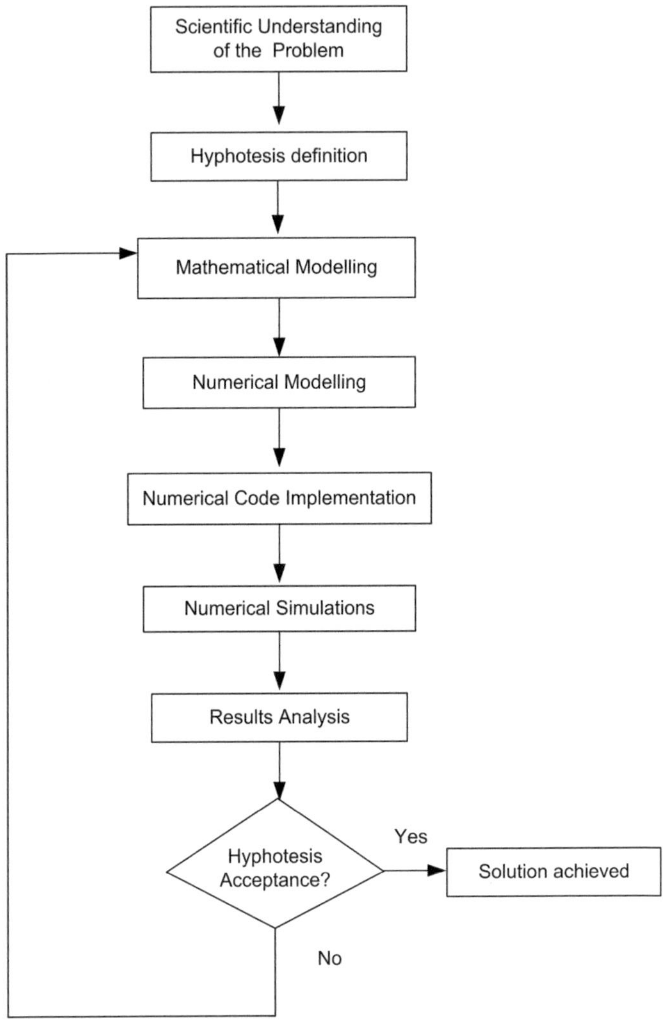

Fig. 1.1 Steps in the scientific method of problems solving

1. Observation and understanding of the problem.
2. Identification of physical law capable to explain the problem (hypothesis definition).
3. Development or use of an existing mathematical model able to describe the physical law.
4. Choice of a numerical method capable of properly solving the equations of the mathematical model with consistency, speed and accuracy.
5. Implementation of a numerical code to perform the numerical simulations.
6. Analysis of the results achieved.
7. Acceptance or rejection of the hypothesis.
   In case of acceptance, the physical-mathematical model provides a solution to the problem. Otherwise, it is necessary to return to Step 2 and repeat the path until a consistent solution is found.

Different approaches can be proposed to solve a problem. As time passes, new models can be proposed (or those already existing improved) in search of a better and more precise solution. This scientific activity is as important as the discovery of the first solution to a problem that troubles us.

The following flowchart shows graphically the steps of the scientific method (Fig. 1.1).

1.3 Objective

This book aims to present the Lagrangian SPH method and its numerical implementation for solving some problems in the area of continuum fluid mechanics and transport phenomena, always taking into account the fundamentals of continuum mechanics.

1.4 Presentation of the Remaining Chapters

Chapter 2 is dedicated to the physical-mathematical modelling of the fluid flows and energy transport, according to the Lagrangian view. The continuum hypothesis and the molecular modelling are presented and discussed. From the use of fundamentals of the continuum mechanics, the physical laws are expressed by means of differential partial equations.

Chapter 3 presents the SPH method fundamentals, boundary treatment techniques, restoration of the consistency and other numerical aspects.

Applications of the SPH method in continuum fluid mechanics and transport phenomena and the analysis of results and discussions on the models implemented and solutions achieved are in Chapter 4.

Appendices bring mathematical deductions related to smoothing functions and the presentation and a general computational SPH code employed in simulations.

At the end of the book, an open-source code is written in FORTRAN programming language for the reader to simulate the heat diffusion in a homogeneous flat plate with Dirichlet boundary conditions.

References

1. Liu, G.R., Liu, M.B.: Smoothed Particle Hydrodynamics: A Meshfree Particle Method. World Scientific, Singapore (2003)
2. Huerta, A., Belytschko, T., Fernández-Mández, S., Rabczuk, T.: Meshfree Methods. Wiley, Encyclopedia of Computational Mechanics (2004)
3. Fox, R.W., McDonald, A.T., Pritchard, P.J.: Introduction to Fluid Mechanics, 6th edn. Wiley, USA (2004)
4. Crespo, A.J.C., Dominguez, J.M., Barreiro, A., Gómez-Gesteira, M., Rogers, B.D.: GPUs, a new tool of acceleration in CFD: Efficiency and reliability on smoothed particle hydrodynamics methods. PLoS ONE **6**(6) (2011). https://doi.org/10.1371/journal.pone.0020685
5. Monaghan, J.J.: Simulating free surface flows with SPH. J. Comput. Phys. **110**, 399–406 (1994)
6. Ataie-Ashtiani, B., Shobeyri, G., Farhadi, L.: Modified incompressible SPH method for simulating free surface problems. Fluid Dyn. Res. **40**, 637–661 (2008)
7. BØckmann, A., Olga, S., Geir, S.: Incompressible SPH for free surface flows. Comput. Fluids (67), 138–151 (2012)
8. Gómez-Gesteira, M., Rogers, B.D., Dalrymple, R.A., Crespo, A.J.C.: State-of-the-art of classical SPH for free-surface flows. J. Hydraul. Res. **48**(Extra Issue), 6–27 (2010)
9. Lo, Y.M.E., Shao, S.: Simulation of near-shore solitary wave mechanics by an incompressible SPH method. Appl. Ocean Res. **24**(5), 275–286 (2002)
10. Narayanaswamy, M., Crespo, A.J.C., Gómez-Gesteira, M., Dalrymple, R.A.: SPHysics-FUNWAVE hybrid model for coastal wave propagation. J. Hydraul. Res. **48**(Extra Issue), 85–93 (2010)
11. Dao, M.H., Xu, H., Chan, E.S., Tkalich, P.: Numerical modelling of extreme waves by smoothed particle hydrodynamics. Nat. Harzards Earth Syst. Sci. **11**, 419–429 (2011)
12. Omidvar, P., Stansby, P.K., Rogers, B.D.: SPH for 3D floating bodies using variable mass particle distribution. Int. J. Numer. Meth. Fluids **72**, 427–452 (2012)
13. Wu, J., Zhang, H., Dalrymple, R.A.: Simulating dam-break flooding with floating objects through intricate city layouts using GPU-based SPH method. In: Proceedings of the World Congress on Engineering—WCE 2013, London, UK (2013)
14. Zhao, X., Hu, C.: Numerical and experimental study on a 2-D floating body under extreme wave conditions. Appl. Ocean Res. **35**, 1–13 (2012)
15. Gómez-Gesteira, M., Dalrymple, R.A.: Using a three-dimensional smoothed particle hydrodynamics method for wave impact on a tall structure. J. Waterw. Port Coast. Ocean Eng. **130**(2), 63–69 (2003)
16. Didier, E., Neves, D.R.C.B., Martins, R., Neves, M.G.: Wave interaction with a vertical wall: SPH numerical and experimental modeling. Ocean Eng. **88**, 330–341 (2014)
17. Altomare, C., Crespo, A.J.C., Domínguez, J.M., Gómez-Gesteira, M., Suzuki, T., Verwaest, T.: Applicability of smoothed particle hydrodynamics for estimation of sea wave impact on coastal structures. Coast. Eng. **96**, 1–12 (2015)
18. Iglesias, A.S., Rojas, L.P., Rodriguez, R.Z.: Simulation of anti-roll tanks and sloshing type problems with smoothed particle hydrodynamics. Ocean Eng. **31**(8–9), 1169–1192 (2004)
19. Colagrossi, A., Lugni, C., Brocchini, M.: A study of violent sloshing wave impacts using an improved SPH method. J. Hydraul. Res. **48**(sup1), 94–104 (2010)
20. Cao, X.Y., Ming, F.R., Zhang, A.M.: Sloshing in a rectangular tank based on SPH simulation. Appl. Ocean Res. **47**, 241–254 (2014)

21. Ghazali, J.N., Kamsin, A.: A real time simulation and modeling of flood hazard. In: 12th WSEAS International Conference on Systems, Heraklion, Greece (2008)
22. Liang, Q., Xia, X., Hou, J.: Efficient urban flood simulation using a GPU-accelerated SPH model. Environ. Earth Sci. **74**, 7285–7294 (2015)
23. Vacondio, R., Rogers, B.D., Stansby, P.K., Mignosa, P.: Shallow water SPH for flooding with dynamic particle coalescing and splitting. Adv. Water Resour. **58**, 10–23 (2013)
24. Cleary, P.W., Prakash, M.: Discrete-element modelling and smoothed particle hydrodynamics: potential in the environmental sciences. Philos. Trans. R. Soc. **362**(1822), 2003–2030 (2004)
25. Cleary, P.W., Prakash, M.: Three dimensional modelling of lava flow using smoothed particle hydrodynamics. Appl. Math. Model. **35**, 3021–3035 (2011)
26. Hérault, A., Bilotta, G., Vicari, A., Rustico, E., Del Negro, C.: Numerical simulation of lava flow using a GPU SPH model. Ann. Geophys. **54**(5), 600–620 (2011)
27. Bilotta, G., Hérault, A., Cappello, A., Ganci, G., Del Negro, C.: GPUSPH: A smoothed particle hydrodynamics model for the thermal and rheological evolution of lava flows. Geol. Soc. Lond. Spec. Publ. **426**, 387–408 (2015)
28. Wang, Z., Chen, R., Wang, H., Liao, Q., Zhu, X., Li, S.: An overview of smoothed particle hydrodynamics for simulating multiphase flow. Appl. Math. Model. **40**, 9625–9655 (2016)
29. Bui, H.H., Sako, K., Fukagawa, R.: Numerical simulation of soil-water interaction using smoothed particle hydrodynamics (SPH) method. J. Terramech. **44**(5), 339–346 (2007)
30. Kristof, P., Benes, B., Krivánek, J., St'ava, O.: Hydraulic erosion using smoothed particle hydrodynamics. Comput. Graph. Forum **28**(2), 219–228 (2009)
31. Wang, C., Wang, Y., Peng, C., Meng, X.: Smoothed particle hydrodynamics simulation of water-soil mixture flows. J. Hydraul. Eng. **142**(10), 04016032 (2016)
32. Manenti, S., Sibilla, S., Gallati, M., Agate, G., Guandalini, R.: SPH simulation of sediment flushing induced by a rapid water flow. J. Hydraul. Eng. **138**, 272–284 (2012)
33. Zou, S.: Coastal sediment transport simulation by smoothed particle hydrodynamics. Phd Thesis, Johns Hopkins University, USA (2007)
34. Tran-Duc, T., Phan-Thien, N., Koo, B.C.: A smoothed particle hydrodynamics (SPH) study of sediment dispersion on the seafloor. Phys. Fluids **29**, 083302 (2017)
35. Shi, H., Yu, X., Dalrymple, R.A.: Development of a two-phase SPH model for sediment laden flows. Comput. Phys. Commun. **221**, 259–272 (2017)
36. Zhu, Y., Fox, P.J., Morris, J.P.: A pore-scale numerical model for flow through porous media. Int. J. Numer. Anal. Meth. Geomech. **23**, 881–904 (1999)
37. Shao, S.: Incompressible SPH flow model for wave interactions with porous media. Coast. Eng. **57**(3), 304–316 (2010)
38. Pereira, G.G., Prakash, M., Cleary, P.W.: SPH modelling of fluid at the grain level in a porous medium. Appl. Math. Model. **35**, 1666–1675 (2011)
39. Tartakovsky, A.M., Trask, N., Pan, K., Jones, B., Pan, W., Williams, J.R.: Smoothed particle hydrodynamics and its applications for multiphase flow and reactive transport in porous media. Comput. Geosci. **20**, 807–834 (2016)
40. Stellingwerf, R.F., Wingate, C.A.: Impact modeling with smooth particle hydrodynamics. Int. J. Impact Eng. **14**, 707–718 (1993)
41. Islam, M.R.I., Chakraborty, S., Shaw, A., Reid, S.: A computational model for failure of ductile material under impact. Int. J. Impact Eng. **108**, 334–437 (2017)
42. LeBlanc, S., Boyer, P., Joslin, C.: Modelling and animation of impact and damage with smoothed particle hydrodynamics. Vis. Comput. **30**, 909–917 (2014)
43. Cleary, P.W., Monaghan, J.J.: Conduction modelling using smoothed particle hydrodynamics. J. Computat. Phys. **148**, 227–264 (1999)
44. Rook, R., Yildiz, M., Dost, S.: Modeling transient heat transfer using SPH and implicit time integration. Numer. Heat Transf. Part B Fundam. **51**(1), 1–23 (2007)

45. Schwaiger, H.F.: An implicit corrected SPH formulation for thermal duffusion with linear free surface boundary conditions. Int. J. Numer. Meth. Eng. **75**, 647–671 (2008)
46. Chen, Y., Kulasegaram, S.: Numerical modelling of fracture of particulate composites using SPH method. Comput. Mater. Sci. **47**, 60–70 (2009)
47. Das, R., Cleary, P.W.: Simulating brittle fracture of rocks using smoothed particle hydrodynamics. In: AIP Conference Proceedings, vol. 1138, p. 1 (2009). https://doi.org/10.1063/1.3155126
48. Douillet-Grellier, T., Jones, B.D., Pramanik, R., Pan, K., Albaiz, A., Williams, J.R.: Mixed-mode fracture modeling with smoothed particle hydrodynamics. Comput. Geotech. **79**, 73–85 (2016)
49. Cleary, P., Joseph, H., Vladimir, A., Nguyen, T.: Flow modelling in casting processes. Appl. Math. Model. **26**, 171–190 (2002)
50. Cleary, P.W., Savage, G., Ha, J., Prakash, M.: Flow analysis and validation of numerical modelling for a thin walled high pressure die casting using SPH. Comput. Part. Mech. **1**, 229–243 (2014)
51. Yi, H., Zhaoyao, Z., Wenjiong, C., Weiping, C.: Simulation of mould filling process using smoothed particle hydrodynamics. Trans. Non-Ferrous Met. Soc. China **21**, 2684–2692 (2011)
52. Shadloo, M.S., Oger, G., Le Touzeb, D.: Smoothed particle hydrodynamics method for fluid flows, towards industrial applications: motivations, current state, and challenges. Comput. Fluids **136**, 11–34 (2016)
53. Leroy, A.: Un nouveau modéle SPH incompressible: vers l'application á des cas industriels. Université Paris-Est, France. http://www.theses.fr/2014PEST1065 (2014). Accessed 20 July 2018
54. Palyanov, A., Sergey, K., Larson, S.D.: Application of smoothed particle hydrodynamics to modeling mechanisms of biological tissue. Adv. Eng. Softw. **98**, 1–11 (2016)
55. Lluch, E., Doste, R., Giffard-Roisin, S., This, A., Sermesant, M., Camara, O., De Craene, M., Morales, H.: Smoothed particle hydrodynamics for electrophysiological modeling: An alternative to finite element methods. In: Proceedings of the 9th International Conference on Functional Imaging and Modelling of the Heart-FIMH 2017, vol. 141, pp. 333–343, Toronto, Canada, Springer International Publishing (2017)
56. Toma, M.: The emerging use of SPH in biomedical applications. Significances Bioeng. Biosci. **1**(1). https://pdfs.semanticscholar.org/bb44/ee090961aff9f80787ffdab66ae2558e6552.pdf (2018). Accessed 20 July 2018
57. Shahriari, S., Kadem, L., Rogers, B.D., Hassan, I.: Smoothed particle hydrodynamics method applied to pulsatile flowinside a rigid two-dimensional model of left heart cavity. Int. J. Numer. J. Numer. Meth. Biomed. Eng. **28**, 1121–1143 (2012)
58. Al-Saad, M., Kulasegaram, S., Bordas, S.: Blood flow simulation using smoothed particle hydrodynamics. In: Proceedings of the VII European Congress on Computational Methods in Applied Sciences and Engineering—ECCOMAS Congress 2016, vol. 4, pp. 8241–8246, Crete Island, Greece. https://www.eccomas2016.org (2016). Accessed 20 July 2018
59. Cleary, P.W., Cohen, R.C.Z., Harrison, S.M., Sinnott, M.D., Prakash, M., Mead, S.: Prediction of industrial, biophysical and extreme geophysical flows using particle methods. Eng. Comput. Int. J. Comput. Aided Eng. Softw. **30**(2), 157–196 (2013)
60. Hieber, S.E., Walther, J.H., Koumoutsakos, P.: Remeshed smoothed particle hydrodynamics simulation of the mechanical behavior of human organs. Technol. Health Care **12**(4), 305–314 (2004)
61. Müller, M., Schirm, S., Teschner, M.: Interactive blood simulation for virtual surgery based on smoothed particle hydrodynamics. Technol. Health Care **12**(1), 25–31 (2004)
62. Ala, G., Francomano, E., Tortorici, A., Toscano, E., Viola, F.: Smoothed particle electromagnetis: a mesh-free solver for transients. J. Comput. Appl. Math. **191**, 194–205 (2006)

63. Ala, G., Francomano, E.: An improved smoothed particle electromagnetics method in 3D time domain simulations. Int. J. Numer. Model. Eletron. Devices Fields **25**, 325–337 (2012)
64. Rundqvist, R., Mark, A., Edelvik, F., Carlson, J.S.: Modeling and simulation of sealing spray application using smoothed particle hydrodynamics. Fluid Dyn. Mater. Process. **7**(3), 259–278 (2011)
65. Dhole, A., Raval, C., Shrivastava, R.: Fluid structure interaction simulation of automotive fuel tank sloshing using nonlinear fluid properties. SAE Technical Paper 2015-26-0240 (2015). https://doi.org/10.4271/2015-26-0240
66. Chiron, L., de Leffe, M., Oger, G., Le Touzé, D.: Fast and accurate SPH modelling of 3D complex wall boundaries in viscous and non viscous flows. Comput. Phys. Commun. (2018). https://doi.org/10.1016/j.cpc.2018.08.001
67. Schnabel, D., Özkaya, E., Biermann, D., Eberhard, P.: Modeling the motion of the cooling lubricant in drilling processes using the finite volume and the smoothed particle hydrodynamics methods. Comput. Methods Appl. Mech. Eng. **329**, 369–395 (2018)
68. Liu, H., Arfaoui, G., Stanic, M., Montigny, L., Jurkschat, T., Lohner, T., Stahl, K.: Numerical modelling of oil distribution and churning gear power losses of gearboxes by smoothed particle hydrodynamics. In: Proceedings of the Institution of Mechanical Engineers, Part J: J. Eng. Tribology (2018). https://doi.org/10.1177/1350650118760626
69. Dong, X.W., Liu, G.R., Li, Z., Zeng, W.: A smoothed particle hydrodynamics (SPH) model for simulating surface erosion by impacts of foreign particles. Tribology International **95**, 267–278 (2016)
70. Smojver, I., Ivancevica, D.: Application of numerical methods in the improvement of safety of aeronautical structures. Transp. Res. Proc. **28**, 164–172 (2017)
71. Xiao, T., Qin, N., Lu, Z., Sun, X., Tong, M.: Development of a smoothed particle hydrodynamics method and its application to aircraft ditching simulations. Aerosp. Sci. Technol. **66**, 28–43 (2017)
72. Lavoie, M.-A., Gakwaya, A., Richard, M.J., Nandlall, D., Zimcik, D.G., Ensan, M.N.: Numerical and experimental modeling for bird and hail impacts on aircraft structure. In: Proceedings of the IMAC-XXVIII, Florida, USA (2010). http://www.am.chalmers.se/~thab/IMAC/2010/PDFs/Papers/s52p002.pdf. Accessed 30 August 2018
73. Goraj, Z.J., Kustron, K.: Review of current research trends in bird strike and hail impact simulations on wing leading edge. Airc. Eng. Aerosp. Tec. **90**(4), 602–612 (2018). https://doi.org/10.1108/AEAT-02-2017-0053
74. Jackson, K.E., Fuchs, W.T.: Comparison of ALE and SPH simulations of vertical drop tests of a composite fuselage section into water. In: Proceedings of the 10th International LS-DYNA Users Conference, Michigan, USA (2008). https://ntrs.nasa.gov/archive/nasa/casi.ntrs.nasa.gov/20080022946.pdf. Accessed 30 August 2018
75. Cartwright, B., Xia, J., Cannon, S., McGuckin, D., Groenenboom, P.: Motion predicition of ships and yachts by smoothed particle hydrodynamics. In: Proceedings of 2nd High Performance Yacht Design Conference, Auckland, New Zealand (2006). http://vm2330.sgvps.net/~syrftest/images/library/20151228144907.pdf. Accessed 30 August 2018
76. Cohen, R.C., Cleary, P.W., Mason, B.R.: Simulations of dolphin kick swimming using smoothed particle hydrodynamics. Hum. Mov. Sci. **31**(3), 604–619 (2012)
77. Monaghan, J.J.: Smoothed particle hydrodynamics and Its diverse applications. Annu. Rev. Fluid Mech. **44**, 323–46 (2012)

2 Physical-Mathematical Modelling

Abstract

This chapter presents the continuum hypothesis, that enables that the three fundamental physical conservation laws (mass, momentum and energy) can be representated by mathematical partial differential equations. Basic concepts that differentiate the microscopic molecular and macroscopic continuum scales are discussed for the first time. Besides, the presentation of the equation of state for the prediction of the dynamic pressure of a fluid flowing, the concept of modified pressure and the modelling of the internal energy is performed in this chapter.

Keywords

Continuum Hypothesis · Physical laws of conservation · Dynamic pressure
Modified pressure

2.1 The Continuum Hyphothesis

When we suppose that the fluid properties vary from point to point in the flow, the continuum hypothesis is being applied [1]. It is the basis of the continuum mechanics. Each point in space is associated to a small volume of fluid that accommodates a large number of molecules in order to make any averaging meaningful. Thus, each spacial point is a material point whose mechanical quantities (density, temperature, viscosity, velocity, acceleration, internal energy, and others) are considered to be functions of position and time and are governed by the classical laws of physics.

In its turn, the molecular modelling investigates the structure, behavior, and physical and mechanical properties of molecular and atomic systems. The methods used can be the quantum mechanics or classical methods. This latter can be deterministic (molecular dynamics) or stochastic (Monte Carlo). In molecular dynamics, potential functions are employed to describe the interatomic forces [2].

C. A. Dutra Fraga Filho, *Smoothed Particle Hydrodynamics*,
https://doi.org/10.1007/978-3-030-00773-7_2

The interaction between atoms and molecules is taken into account by the molecular modelling, but it is not considered by the continuum mechanics, which does not have the capability to represent the atoms and their bonds. The continuum medium, in which the molecular motion and the intermolecular forces are not considered, is studied macroscopically. The absence of voids in space allows the use of differential mathematics as a tool for writing the classical physical laws of conservation of mass, momentum and energy.

2.2 Physical Laws of Conservation

The mathematical modelling of a fluid flow and energy transport is performed by the mass, momentum, and energy laws, which are expressed by the conservation equations. The fluid flow is described by the velocity fields, density fields and pressure fields. The distributions of energy are described by the temperature fields.

The evolution of the density, velocity and energy fields over time are defined, according to the Lagrangian view, by Eqs. (2.1), (2.2) and (2.3)—conservation of mass, momentum and energy, respectively, as follows:

$$\boxed{\frac{D\rho}{Dt} + \rho\nabla\cdot\vec{\mathrm{v}} = 0} \tag{2.1}$$

where

t is the time
$\frac{D}{Dt}$ is the Lagrangian (or material) derivative
ρ is the density
$\vec{\mathrm{v}}$ is the velocity
∇ is the mathematical nabla operator, a fictional vector of partial derivatives $\partial/\partial\mathrm{x}$ with respect to x, $\partial/\partial\mathrm{y}$ with respect to y and $\partial/\partial\mathrm{z}$ with respect to z.

$$\boxed{\rho\frac{D\vec{\mathrm{v}}}{Dt} = -\nabla P + \mu\nabla^2\vec{\mathrm{v}} - \frac{2}{3}\nabla\left(\mu\nabla\cdot\vec{\mathrm{v}}\right) + \vec{f}_{ext}} \tag{2.2}$$

where

P is the absolute pressure
μ is the absolute viscosity
$\mu\nabla^2\vec{\mathrm{v}}$ are the viscosity forces per unit volume
$\vec{f}_{ext}$ are the external forces acting on the fluid, per unit volume
∇P are the pressure forces per unit volume.

In this book, the only external force, per unit volume, acting on the fluid is $(\rho \vec{g})$, in which $\vec{g}$ is the gravity.

$$\boxed{\rho \frac{De}{Dt} = -P(\nabla \cdot \vec{\mathrm{v}}) + \varepsilon_{\mathrm{v}} + \nabla \cdot (K \nabla T) + \dot{q}} \tag{2.3}$$

where

e	is the internal specific energy
ε_{v}	is the energy dissipation rate per unit volume
K	is the thermal conductivity
T	is the absolute temperature
$P(\nabla \cdot \vec{\mathrm{v}})$	is the work rate due to pressure forces, per unit volume
$\nabla \cdot (K\nabla T)$	is the heat rate due to the diffusion, per unit volume
$\dot{q}$	is the heat rate provided by the source, per unit volume.

The equations above are written for a Newtonian fluid. Analysing the conservation equations in a 3-D domain:

- The law of conservation of mass presents only one unknown: the density.
- The momentum conservation law is composed of three equations (one in each Cartesian direction), presenting four unknowns: the three components of velocity and the pressure (absolute viscosity and acceleration of gravity are considered known and constant). In order to be solved, a state equation is used to calculate the pressure, presented in the next section.
- There are two unknowns in the energy conservation law: the specific energy and the temperature. The thermal conductivity, energy dissipation rate per unit volume and the heat generated by the sources per unit volume are considered constant and known. This way, we have an equation and two unknowns. The specific internal energy must be modelled so that Eq. (2.3) can be solved.

2.3 Pressure Modelling

2.3.1 Equation of State for Dynamic Pressure

In fluid dynamics, the compressible fluid can be approximated to an incompressible fluid by means of quasi-compressible fluid modelling. For free surface flows, the dynamic pressure can be calculated through an equation of state, known as the Tait equation:

$$\boxed{P_{dyn} = B\left(\left(\frac{\rho}{\rho_o}\right)^{\gamma} - 1\right)} \tag{2.4}$$

where

P_{dyn}	is the dynamic pressure
B	is the term associated with fluctuations in the fluid density
ρ_o	is the fluid density at rest
Ma	is the Mach number
$\gamma = 7$	is a reference value, used in most circumstances.[1]

The term B can be estimated by:

$$B = \frac{c^2 \rho_o}{\gamma} \tag{2.5}$$

where c is the magnitude of the sound velocity in the fluid.

In most cases, B can be taken as the initial pressure [4].

In order for the Tait equation to be applied to the prediction of the fluid pressure field, the maximum value of the Mach number should be 0.1.

$$Ma = \frac{\mathrm{v}}{c} \leq 0.1 \tag{2.6}$$

where v is the magnitude of the flow velocity.

A second choice of the equation of state is:

$$\boxed{P_{dyn} = c^2 \rho} \tag{2.7}$$

Liu and Liu [4] simulated Couette and Poiseuille flows using this equation of state, and good results have been obtained.

In the fluid flow, the absolute pressure is composed by two parcels: the hydrostatic pressure (due to the fluid column over the point of the flow considered) and the the dynamic pressure predicted by the equation of state.

The following section presents the concept of modified pressure, which can be useful in solving certain problems.

2.3.2 Modified Pressure

Batchelor [5] introduced the modified pressure concept, defined as the remaining part of the pressure that exceeds the parcel sufficient to balance the action of the gravitational force, causing the fluid motion. Equation (2.8) presents modified pressure in a mathematic form:

$$\boxed{P_{\mathrm{mod}} = (P - P_o) - \rho g(H - \mathrm{y})} \tag{2.8}$$

[1] For lower Reynolds numbers, more accurate pressure estimates are obtained if γ is equal to 1, since errors in density and pressure remain proportional [3].

where

P_{mod}	is the modified pressure
H	is the ordinate of the fluid surface above the point analysed
P_o	is the reference pressure
y	is the ordinate of the position occupied by the fluid.

For an incompressible fluid ($\nabla \cdot \vec{v} = 0$) the momentum equation is written as follows:

$$\boxed{\rho \frac{D\vec{v}}{Dt} = -\nabla P_{mod} + \mu \nabla^2 \vec{v}} \tag{2.9}$$

It is seen that, employing the modified pressure, the gravity disappears from the equation of motion.

In hydrostatic situations the modified pressure is zero at all points of the fluid. In dynamic cases, it is the difference between the dynamic pressure of the fluid and the term $\rho g(H - y)$.

In general, the modified pressure is applied to flows without free surfaces. Gravity will reenter the problem through boundary conditions if it is, in fact, a significant driving force for fluid motion [6].

2.4 Specific Internal Energy Modelling

For incompressible fluid flows, specific heat at constant volume (c_v) is defined as follows:

$$c_v \equiv \left(\frac{\partial e}{\partial T}\right)_v \tag{2.10}$$

Observing the previous consideration (constant volume), the specific internal energy is only a function of the temperature, which leads to the relation:

$$c_v = \frac{De}{DT} \Rightarrow De = c_v DT \tag{2.11}$$

References

1. Tritton, D.J.: Physical Fluid Dynamics, 2nd edn. Oxford Science Publications, New York (1988)
2. Rapaport, D.C.: The Art of Molecular Dynamics Simulation, 3rd edn. Cambridge University Press, UK (2004)
3. Morris, J.P., Fox, P.J., Zhu, Y.: Modeling low Reynolds number incompressible flows using SPH. J. Comput. Phys. **136**, 214–226 (1997)

4. Liu, G.R., Liu, M.B.: Smoothed Particle Hydrodynamics: A Meshfree Particle Method. World Scientific, Singapore (2003)
5. Batchelor, G.K.: An Introduction to Fluid Dynamics, 3rd edn. Cambridge University Press, UK (2000)
6. Clark, M.M.: Transport Modeling for Environmental Engineers and Scientists, 2nd edn. Wiley, New Jersey (2000)

3 Smoothed Particle Hydrodynamics Method

Abstract

In this chapter, the basic concepts of the SPH method and its application in the discretization of the continuum domain in particles are presented. The SPH approximations of the equations of conservation are deduced and explained. Kernels used in interpolations, temporal integration methods, the particle inconsistency problem and numerical corrections applied in simulations are also presented. A special attention is given to the presentation of common boundary condition techniques applied in SPH. The use of fictitious (virtual, ghost, mirror or dummy) particles, with or without artificial repulsive forces, or dynamic particles is discussed, and boundary conditions techniques which respect the laws of physics and continuum mechanics, as the reflective boundary condition are defended.

Keywords

SPH fundamentals · Domain discretization · Conservation equations
Free surface · Interfaces · Numerical aspects · Consistency · Boundary conditions

3.1 Fundamentals

Originally, the SPH method was developed in the end of the 1970s, by [1] and [2], for the modelling of astrophysics phenomena. After some time, it was employed in the field of solid and fluid mechanics with a wide range of applications, due to its ability to incorporate the complexity of physical problems.

The essence of this Lagrangian method is to discretize the continuum domain into a finite number of particles which get the physical quantities of interest from weighted interpolations of the quantities in the neighbouring particles. Only those particles that are within the domain of influence (at a maximum distance kh from the

C. A. Dutra Fraga Filho, *Smoothed Particle Hydrodynamics*,
https://doi.org/10.1007/978-3-030-00773-7_3

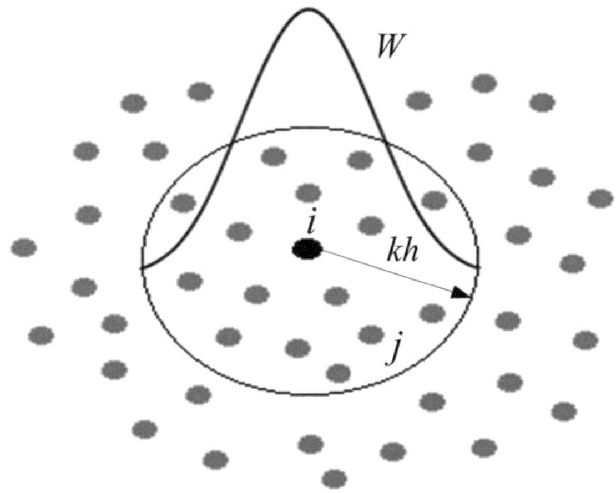

Fig. 3.1 Graphical representation of the kernel and its domain of influence. The reference particle '*i*' (black central circle) has its physical properties influenced by the neighbour particles '*j*', which are within the domain of influence

fixed particle considered) will contribute to the behaviour of the physical quantities in the fixed particle.

Figure 3.1 shows a graphical representation of the Lagrangian particles within the domain of influence (circumference of radius equal to kh).

SPH is based on the mathematical identity, valid for a defined and continuous function $f(X)$, as shown the equation below:

$$f(X) = \int_{\Omega} f(X')\delta(X - X')dX' \tag{3.12}$$

$$\delta(-X') = \begin{cases} 1, \text{ if} X = X', \\ 0, \text{in the other case.} \end{cases} \tag{3.13}$$

where

$X = (\text{x}, \text{y}, \text{z})$ is the position of the fixed point
$X' = (\text{x}', \text{y}', \text{z}')$ is the position of the variable point
$f(X)$ is the value of the scalar function at the fixed point
$f(X')$ is the value of the scalar function at the variable point
$\delta(X - X')$ is the Dirac delta function evaluated in the position $(X - X')$
Ω is the domain
dX' is the infinitesimal volume element.

In the notation used in this book, X is a point in space determined by the coordinates x, y and z; that is, $X = (\text{x}, \text{y}, \text{z}) = \text{x}\mathbf{i} + \text{y}\mathbf{j} + \text{z}\mathbf{k}$ (where **i**, **j** and **k** are the unit vectors in a 3-D Euclidean space). Throughout the text, a point in space can be seen as a vector.

By replacing the Dirac delta function by a kernel or smoothing function (W), the approximation to the function f in position X is obtained, resulting in Eq. (3.14):

$$< f(X) > = \int_{\Omega} f(X')W(X - X', h)dX' \tag{3.14}$$

where
$< f(X) >$ is the approximate value of the scalar function f at the fixed point X
h is the smoothing length
$W(X - X', h)$ is the kernel evaluated in the position $(X - X')$.

In the SPH method, the smoothing function (kernel) ensures a greater contribution from the nearest neighbouring particles for the value of the physical property in the reference particle. The kernel satisfies some properties:

- **Positivity**: $W(X - X', h) \geq 0$.
- **Symmetry**: $W(X - X', h) = W(X' - X, h)$.
- **Unity (normalized over the domain of influence)**: $\int_{\Omega} W(X - X', h)dX' = 1$.
- **Compact support**: $W(X - X', h) = 0$, when $|X - X'| > kh$, where kh is the support radius of the domain of influence and k is a scaling factor (that determines the spread of the smoothing function).
- **Delta function property (convergency)**: $\lim_{h \to 0} W(X - X', h) = \delta(X - X')$.
  The kernel must satisfy the Dirac delta function as the distance between the fixed and variable points approaches zero.
- **Decay**: Its value decreases monotonically as the assessed point moves away from the central point of the domain of influence.
- **Smoothness**: The smoothing function should be sufficiently smooth.

Any function that respects the properties above can be used as a smoothing function in the SPH method.

3.2 Discretization of the Continuum Domain

After the discretization of the continuum domain, a set of differential partial equations with respect to time is obtained for transient problems. A particle method has the purpose of providing a solution to these equations that express the laws of physics.

The SPH method employs particles to solve problems in continuum mechanics. The two approaches, particle and continuum, are in agreement since the microscopic molecular fluctuations, which are absent in continuum mechanics, can be ignored.

Discretizing the continuum domain using particles, the value of the function f will be known in some points of the domain (positions of the mass centres of the Lagrangian elements). The infinitesimal volume element is defined as $dX' = (m'/\rho')$, in which m' is the mass and ρ' is the density of the infinitesimal volume element of the fluid.

Equation (3.14) written after particle discretization leads to the general expression for the SPH method (Eq. 3.15):

$$\boxed{f_i = \sum_{j=1}^{n} m_j \frac{f_j}{\rho_j} W(X_i - X_j, h)} \tag{3.15}$$

where
f_i is the approximate value of the function in the position of the fixed particle
f_j is the approximate value of the function in the position of the neighbouring particle
m_j is the mass of the neighbouring particle
ρ_j is the density of the neighbouring particle
X_i is the position of the fixed particle
X_j is the position of the neighbouring particle
$W(X_i - X_j, h)$ is the kernel evaluated at the position $(X_i - X_j)$
n is the number of neighbouring particles.

Due to the compact support property, the summation is limited to the number of neighbouring particles, not to the entire computational domain.

The success in the approximations also depends on the existence of a sufficient and adequate number of particles within the domain of influence (defined by the circumference of radius kh) and also by an appropriate length of this radius. The properties of the kernel (presented in the previous section) and its gradient (anti-symmetry and non-normalization, in the Euclidean coordinate system, discussed in Appendix 4.2.2) must be assured, and, consequently, the accuracy of results is obtained in interpolations.

For the interpolations to be representative in the discretized domain, the minimum number of neighbouring particles within the domain of influence should be 5, 21 and 57 in cases 1-D, 2-D and 3-D, respectively, [3]. After computational tests have been performed, it has been seen that increasing the number of neighbouring particles further does not significantly alter the numerical results, but does, however, affect the time spent in the simulation. Figure 3.2 shows a 2-D cell with 21 particles within the domain of influence.

The radius of the domain of influence, kh, also influences the results of the interpolations. Therefore, its choice needs to be properly analysed [4].

3.2.1 Approximation of the Divergent of a Vectorial Function

In order to obtain an SPH approximation of the divergent of a vectorial function **f**, we substitute $f(X)$ by $\nabla.\mathbf{f}(X)$, in Eq. (3.14):

$$< \nabla.\mathbf{f}(X) >= \int_{\Omega} [\nabla.\mathbf{f}(X')] W(X - X', h) dX' \tag{3.16}$$

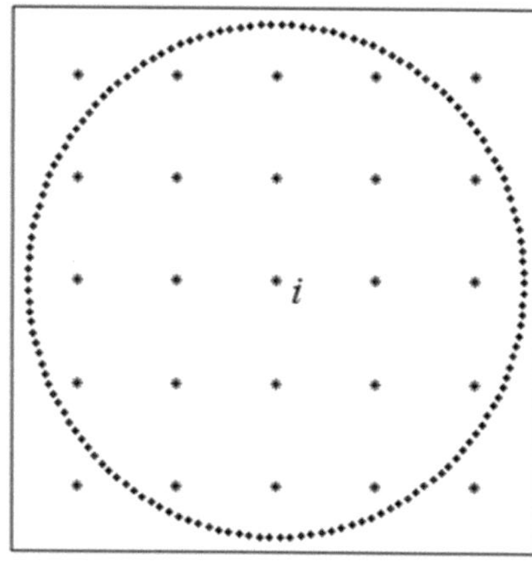

Fig. 3.2 A two-dimensional cell with 21 particles within the domain of influence

where
$\nabla.\mathbf{f}(X)$ is the approximation for the divergent of the vectorial function at the position X
$\mathbf{f}(X')$ is the vectorial function evaluated at the position of the variable point.

It is known that:

$$\left[\nabla.\mathbf{f}(X')\right] W(X-X',h) = \nabla.\left[\mathbf{f}(X')W(X-X',h)\right] - \mathbf{f}(X').\nabla W(X-X',h) \tag{3.17}$$

Combining Eqs. (3.14) and (3.17):

$$< \nabla.\mathbf{f}(X) >= \int\limits_{\Omega} \nabla.\left[\mathbf{f}(X')W(X-X',h)\right]dX' - \int\limits_{\Omega} \mathbf{f}(X').\nabla W(X-X',h)dX' \tag{3.18}$$

Applying the divergence (or Gaussian) theorem to the first term on the right-hand side of the previous equation:

$$< \nabla.\mathbf{f}(X) >= \int\limits_{S} \mathbf{f}(X')W(X-X',h)\mathbf{n}.d\mathbf{S} - \int\limits_{\Omega} \mathbf{f}(X').\nabla W(X-X',h)dX' \tag{3.19}$$

where
S is the integration surface
$d\mathbf{S}$ is the infinitesimal surface element
$\mathbf{n}$ is the unit vector normal to the surface.

Due to the kernel's compact support property, we have:

$$< \nabla.\mathbf{f}(X) >= -\int_{\Omega} \mathbf{f}(X').\nabla W(X - X', h)dX' \tag{3.20}$$

And, after the discretization by particles:

$$\nabla.\mathbf{f}_i = -\sum_{j=1}^{n} \mathbf{f}_j.\nabla W(X_i - X_j, h)\frac{m_j}{\rho_j} \tag{3.21}$$

where
$\nabla.\mathbf{f}_i$ is the divergent of the vectorial function in the position occupied by the fixed particle
$\mathbf{f}_j$ is the vectorial function evaluated at the position occupied by the neighbouring particle.

This is the basic approximation form of the divergent of a vectorial function in the SPH method.

The use of the approximation in its basic form does not lead to the best results in fluid flow applications. Monaghan [5] obtained a second form for the approximation of the divergent from the use of mathematical identity:

$$\nabla.\mathbf{f} = \frac{1}{\rho}[\nabla.(\rho\mathbf{f}) - \mathbf{f}.\nabla\rho] \tag{3.22}$$

This leads to a new approach of the divergent of a function:

$$\boxed{\nabla.\mathbf{f}_i = \frac{1}{\rho_i}\sum_{j=1}^{n}(\mathbf{f}_j - \mathbf{f}_i).\nabla W(X_i - X_j, h)m_j} \tag{3.23}$$

where
$\mathbf{f}_i$ is the vectorial function evaluated at the position occupied by the fixed particle
ρ_i is the density of the fixed particle.

3.2.2 Approximation of the Gradient of a Scalar Function

Through a procedure analogous to that achieved in obtaining the approximation of the divergent of a vector function, the approximation of the gradient of a scalar function f is achieved:

$$< \nabla f(X) >= -\int_{\Omega} f(X')\nabla W(X - X', h)dX' \tag{3.24}$$

where $< \nabla f(X) >$ is the approximation of the gradient at the position X

In the discretized form, used in the SPH method, we have:

$$\nabla f_i = -\sum_{j=1}^{n} f_j \nabla W(X_i - X_j, h) \frac{m_j}{\rho_j} \tag{3.25}$$

where ∇f_i is the gradient of the function evaluated in the position occupied by the fixed particle

This is the basic form of the approximation of the gradient of a function in SPH.

There are other more accurate forms of this approximation, as regards the obedience to conservation equations. The gradient of a scalar function can be written using the rule of the derivative of the product [5]:

$$\nabla f = \frac{1}{\rho} [\nabla(\rho f) - f \nabla \rho] \tag{3.26}$$

Employing Eq. (3.25), we have:

$$\boxed{\nabla f_i = -\frac{1}{\rho_i} \sum_{j=1}^{n} m_j (f_j - f_i) \nabla W(X_i - X_j, h)} \tag{3.27}$$

In situations in which the interaction between two particles must be symmetric, a third form has been employed to approximate the gradient of a function. This occurs in cases, such as the approximation of the pressure gradient, in which two particles, occupying different positions in space, have different pressures at a certain instant of time. When using the symmetrical approximation, as shown below, it is ensured that the forces exerted between a pair of particles have the same magnitude. Starting from the mathematical identity:

$$\nabla \left(\frac{f(X)}{\rho} \right) = \frac{\nabla f(X)}{\rho} + \frac{f(X)}{\rho^2} \nabla \rho \quad \Rightarrow \tag{3.28}$$

$$\frac{\nabla f(X)}{\rho} = \nabla \left(\frac{f(X)}{\rho} \right) - \frac{f(X)}{\rho^2} \nabla \rho \tag{3.29}$$

And obtaining the SPH approximations of the two terms on the right-hand side of the previous equation (from Eq. 3.25):

$$\nabla \left(\frac{f(X)}{\rho} \right)_i = -\sum_{j=1}^{n} m_j \left(\frac{f_j}{\rho_j{}^2} \right) \nabla W(X_i - X_j, h) \tag{3.30}$$

$$\left(\frac{f(X)}{\rho^2} \nabla \rho \right)_i = -\sum_{j=1}^{n} \frac{f_i}{\rho_i^2} m_j \nabla W(X_i - X_j, h) \tag{3.31}$$

Finally, replacing Eqs. (3.30) and (3.31) in Eq. (3.29), we obtain the expression for the symmetric approximation of the gradient, in a Lagrangian particle 'i':

$$\boxed{\nabla f_i = -\rho_i \sum_{j=1}^{n} m_j \left(\frac{f_i}{\rho_i^2} + \frac{f_j}{{\rho_j}^2} \right) \nabla W(X_i - X_j, h)} \tag{3.32}$$

The symmetrical form of the approximation of the gradient was the one used in the studies presented in this book.

3.2.3 Approximation of the Laplacian

The Laplacian approximation can be obtained from the expansion of the Taylor series around a fixed spatial point.

In the Cartesian coordinate system, the expression for the Laplacian of a function has the following form:

$$\boxed{\Delta f_i = \left(\frac{\partial^2 f_i}{\partial \mathrm{x}^2} + \frac{\partial^2 f_i}{\partial \mathrm{y}^2} \right) = 2 \sum_{j=1}^{n} \frac{m_j}{\rho_j} \left(f_i - f_j \right) \Delta X_{ij} . \nabla W \left(X_i - X_j, h \right)} \tag{3.33}$$

$$\Delta X_{ij} = \frac{\mathrm{X}_i - \mathrm{X}_j}{\left| \mathrm{X}_i - \mathrm{X}_j \right|^2} \tag{3.34}$$

where Δf_i is the SPH approximation of the Laplacian of the function at the position of the fixed particle.

In the polar coordinates system, the expression of the SPH Laplacian of a function becomes:

$$\boxed{\Delta f_i = 2 \sum_{j=1}^{n} \frac{m_j}{\rho_j} \left(f_i - f_j \right) \frac{\partial W(\mathbf{r}_i - \mathbf{r}_j, h)}{\partial r} \frac{1}{\left| \mathbf{r}_{ij} \right|}} \tag{3.35}$$

where
r is the radial direction
$W(\mathbf{r}_i - \mathbf{r}_j, h)$ is the smoothing function written in polar coordinates, evaluated at the position $(\mathbf{r}_i - \mathbf{r}_j)$
$\mathbf{r}_{ij} = (\mathbf{r}_i - \mathbf{r}_j)$ is the radial distance between the fixed and a neighbouring particle.

The mathematical deduction of the Laplacian approximation is presented in detail in the Appendix B.

Once obtained the approximations for the divergent, gradient and Laplacian of a function, it is also possible to obtain SPH approximations for the conservation equations.

3.2.4 SPH Approximations to the Conservation Equations

3.2.4.1 Mass Conservation

The classic form of the SPH approximation to the density field, known as summation approach, is obtained by the substitution of f for ρ in the continuity Eq. (3.15):

$$\boxed{\rho_i = \sum_{j=1}^{n} m_j W(X_i - X_j, h)} \tag{3.36}$$

From the above equation, it is concluded that the kernel's unit is the inverse unit of volume.

One alternative form, also much used in the approximation of the density field, is obtained from the equation of conservation of mass (Eq. 2.1):

$$\boxed{\frac{D\rho_i}{\mathrm{Dt}} = \sum_{j=1}^{n} m_j \left(\vec{\mathrm{v}}_i - \vec{\mathrm{v}}_j\right).\nabla W(X_i - X_j, h)} \tag{3.37}$$

where
$\vec{\mathrm{v}}_i$ is the velocity of the fixed particle
$\vec{\mathrm{v}}_j$ is the velocity of the neighbouring particle.

3.2.4.2 Momentum Conservation

$$\boxed{\begin{aligned}\frac{\mathrm{D}\vec{\mathrm{v}}_i}{Dt} &= \sum_{j=1}^{n} m_j \left[\frac{P_i}{(\rho_i)^2} + \frac{P_j}{(\rho_j)^2}\right] \nabla W(X_i - X_j, h) + \\ &2\upsilon_i \sum_{j=1}^{n} \frac{m_j}{\rho_j} \left(\vec{\mathrm{v}}_i - \vec{\mathrm{v}}_j\right) \frac{(X_i - X_j)}{\left|X_i - X_j\right|^2}.\nabla W(X_i - X_j, h) + \vec{g}\end{aligned}} \tag{3.38}$$

where
P_i is the absolute pressure acting on the fixed particle
P_j is the absolute pressure acting on the neighbouring particle
υ_i is the kinematic viscosity of the fixed particle.

3.2.4.3 Energy Conservation

$$\frac{De_i}{Dt} = -P_i \sum_{j=1}^{n} m_j \left(\vec{v}_i - \vec{v}_j\right).\nabla W(X_i - X_j, h) + \varepsilon_v + \nabla.\mathbf{q} + q_H \tag{3.39}$$

where e_i is the specific internal energy of the fixed particle.

3.2.5 Errors in SPH Approximations

The approximations to a function and its derivatives given by the SPH method have errors. This subject will be discussed below.

Applying the expansion of the Taylor series to the function f, which is differentiable at the domain, around the fixed point X:

$$< f(X) >= \int_{\Omega} \left[f(X') + f'(X)(X' - X) + \mathrm{Re}(X' - X)\right] W(X - X', h)dX' \tag{3.40}$$

$$< f(X) >= \int_{\Omega} f(X')W(X - X', h)dX' + f'(X) \int_{\Omega} (X' - X)W(X - X', h)dX' + \mathrm{Re}(h^2) \tag{3.41}$$

where

$f'(X)$ is the first-order derivative of the function evaluated in the position X

$\mathrm{Re}(h^2)$ is the residue of second order, after the truncation of the Taylor series.

The kernel is an even function. Due to this, the expression shown in Eq. (3.14) of the approximation of the function f is obtained. That is, there is a residue of second order, arising from the truncation of the Taylor series, as shown in this subsection, when Eq. (3.14) is used to obtain the approximation of a function by the SPH method. Liu and Liu [4], Quinlan et al. [6] and Vaughan et al. [7] are references that present good studies about this matter.

3.2.6 Smoothing Functions

The smoothing functions must obey the properties presented in Sect. 3.1. Furthermore, the smoothing function must be twice continuously differentiable and both the kernel and its derivatives computationally inexpensive.

There are many possible choices of the smoothing function. However, the kernel used influences in the achievement of best approximations of the physical properties of the particles. The cubic spline kernel is widely used due to its mathematical behaviour, as well as its derivatives [4].

- Cubic Spline Kernel:

$$W(X-X',h)=\alpha_D\begin{cases}\left(\frac{2}{3}-q^2+\frac{1}{2}q^3\right),0\leq q\leq 1\\ \left(\frac{1}{6}(2-q)^3\right),1<q\leq 2\\ 0,\text{ in the other case.}\end{cases} \tag{3.42}$$

where
$\alpha_D=\frac{15}{7\pi h^2}$ and $\frac{3}{2\pi h^3}$ in 2-D and 3-D domains, respectively.

Other kernels mcan be employed in the SPH method, such as the quartic proposed by Lucy [1], the new quartic presented by Liu and Liu [4] and the quintic spline proposed by Morris et al. [8], presented below.

- Lucy's Quartic Kernel:

$$W(X-X',h)=\alpha_D\begin{cases}(1+3q)\left(1-q^3\right),0\leq q\leq 1\\ 0,\text{ in the other case.}\end{cases} \tag{3.43}$$

where
$q=\frac{|(X-X',h)|}{h}$
α_D is a constant that assures the kernel normalization over the domain of influence, and it assumes the values $\frac{5}{\pi h^2}$ (in 2-D domains) and $\frac{105}{16\pi h^3}$ (in 3-D domains).

- New Quartic Kernel:

$$W(X-X',h)=\alpha_D\begin{cases}\left(\frac{2}{3}-\frac{9}{8}q^2+\frac{19}{24}q^3-\frac{5}{32}q^4\right),0\leq q\leq 2\\ 0,\text{ in the other case.}\end{cases} \tag{3.44}$$

where
$\alpha_D=\frac{15}{7\pi h^2}$ and $\frac{315}{208\pi h^3}$ in 2-D and 3-D domains, respectively.

- Quintic Spline Kernel:

$$W(X-X',h)=\alpha_D\begin{cases}(3-q)^5-6(2-q)^5+15(1-q)^5,0\leq q\leq 1\\ (3-q)^5-6(2-q)^5,1<q\leq 2\\ (3-q)^5,2<q\leq 3\\ 0,\text{ in the other case.}\end{cases} \tag{3.45}$$

where
$\alpha_D = \dfrac{7}{478\pi h^2}$ and $\dfrac{1}{120\pi h^3}$ in 2-D and 3-D domains, respectively.

The mathematical procedure of obtaining the constant α_D is presented in Appendix A.

Liu et al. [9], Hongbin and Xin [10], Dehnen and Hossam [11] and Yang et al. [12] are interesting literature references on smoothing functions employed in SPH interpolations.

3.2.7 Neighbouring Particles Search

The literature on Lagrangian methods presents several techniques for the search of neighbour particles (a necessary procedure to be performed, especially in fluid flows, at each numerical iteration, and that has a direct influence on the simulation time). In two-dimensional analyses, neighbour lists (linked and Verlet) are commonly utilized in particle simulations [13, 14]. In 3-D problems, the octree technique [15, 16] is being currently applied in conjunction with parallelization, or CUDA GPU processing.

In the direct search, the simplest searching method, a particle '*i*', is fixed and all pairs of particles '*ij*' in the domain will have their distances calculated and compared to the value of kh. If the distance is smaller than the support radius kh, a neighbour particle is found and storaged in a matrix. Figure 3.3 shows the regions where the search is performed in the direct search, and using a grid (in which only the particles within the reduced light blue region will be analysed).

Grids are used in neighbour lists techniques, as linked and Verlet. Viccione et al. [14] built a cell-linked list by partitioning the physical space with a regular grid of congruent rectangles (2-D domain) or rectilinear parallelepipeds (3-D). Particles are simply assigned to cells according to their spatial coordinates. Each cell contains a number of particles that can vary during the numerical simulation. Figure 3.4 presents a flow chart of a simple linked list code.

According to Gingold and Monaghan [17], if the particles are assigned to cells and identified through linked lists, the calculation time is proportional to N and not N^2 as it would be if the complete summation was performed.

After studies were carried out to verify the computational efficiency of the linked list, it was verified that the simple substitution of the direct search by this technique does not result in smaller search times. On the other hand, according to Liu and Liu [4], the storage of the neighbouring particles in pairs leads to an improvement in the search process employing the linked list. This statement proved to be true, with a significant reduction in computational time used by the linked list technique.

The Verlet list is a linked list optimization proposal in which the neighbour list is not updated at each numerical iteration. In this technique, the neighbour list is built with probable neighbouring particles of each reference particle at domain, inside a cut-off radius $(kh + L)$ slightly higher than the support radius kh used in the particle method. The criterion for the definition of the cut-off radius is based on the maximum particle displacement during the time in which the neighbour list will remain

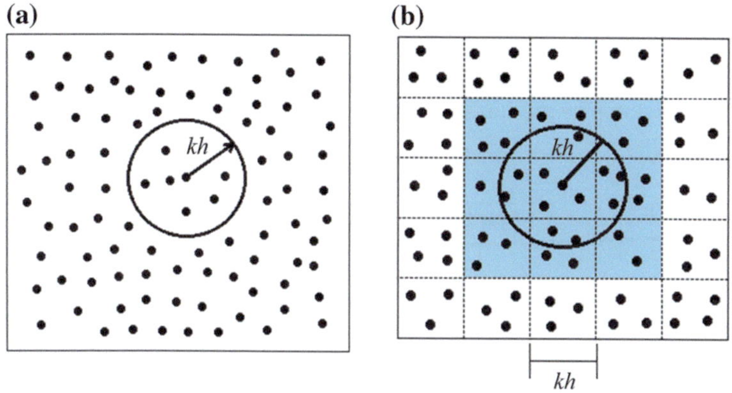

Fig. 3.3 The search for neighbouring particles: **a** directly and **b** using a grid

the same (which is obtained from the maximum velocity of the particle and simulation time in which the list will not be updated).

Figure 3.5 shows all particles within the cut-off radius $(kh + L)$ that will be added to a list of potential neighbour particles. However, only those particles whose distances for the reference particle are less than or equal to kh will be used in the interpolations of the particle method. The choice of the cell size must be done carefully in order to guarantee the accuracy and the advantages of the technique. The neighbour list is not updated at numerical iteration. So the processing time savings is expected.

From the literature results [13], it is concluded that a wise choice of some parameters, such as cell size and number of time steps in which the neighbour list will remain unchanged, is necessary so that a reasonable improvement in the processing time (below 10 %) is reached.

3.2.8 Treatment of the Free Surface

In the Lagrangian treatment, the free surface is identified by the particles themselves, by their positions in space at a specific instant of time. On the free surface, there is an imbalance of forces acting on the particles that are therein. These force must be taken into account in the physical-mathematical modelling of the problem. Normally, they are not treated by the Navier–Stokes equations because they are considered as boundary conditions.

Different methods are used for the calculation of the force of surface tension. Here, two of the various approaches available in the literature.

3.2.8.1 Continuum Surface Force Method

The continuum surface force (CSF) method was developed by Brackbill et al. [18]. The force generated by the surface tension acting on a particle on the surface of the fluid is given by:

$$\boxed{\mathbf{fs}(X_s) = \sigma\kappa\mathbf{n}(\mathrm{X_S}) + \nabla_s\sigma} \tag{3.46}$$

where
$\mathbf{fs}(X_s)$ is the force of surface tension acting in the direction normal to the free surface, at the point of free surface (X_s)
σ is the fluid surface tension coefficient (a constant which varies from fluid to fluid), in units of force per unit length
κ is the curvature of the surface
$\mathbf{n}(\mathrm{X_S})$ is the vector normal to the free surface, located at the point X_s, in the interface
∇_s is the gradient of the surface.

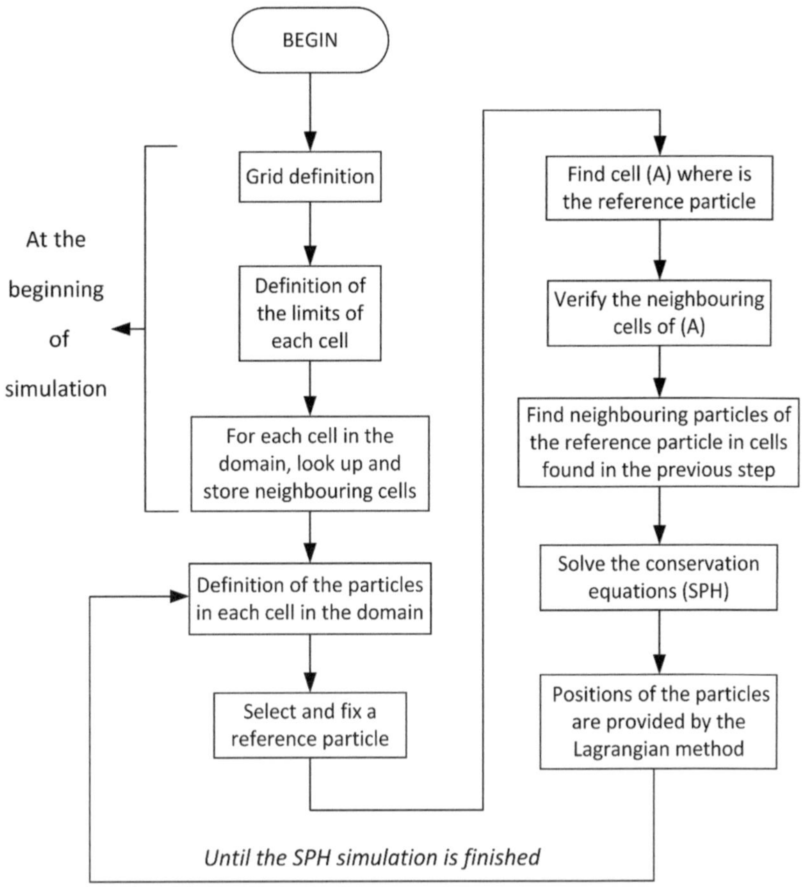

Fig. 3.4 Flow chart showing the operations performed in a linked list code

The first term in equation above is a force which acts in the direction normal to the free surface. This force smooths regions with high curvature and works to reduce the total surface area. Figure 3.6 shows the direction of force of surface tension acting on different points of the surface.

The second term of Eq. (3.46) is a force which acts along a tangent of the interface. This force works to move fluid from regions of low surface tension to higher surface tension. Considering that the surface tension is constant throughout the fluid, the second term can be ignored. The force to be calculated reduces to:

$$\boxed{\mathbf{fs}_i = \sigma\kappa\mathbf{n}_\mathrm{i}} \tag{3.47}$$

where

$\mathbf{fs}_i$ is the force of surface tension exerted in the position of the particle surface i, in the direction normal to the interface

$\mathbf{n}_\mathrm{i}$ is the vector normal to the surface, located in the position of the particle surface i.

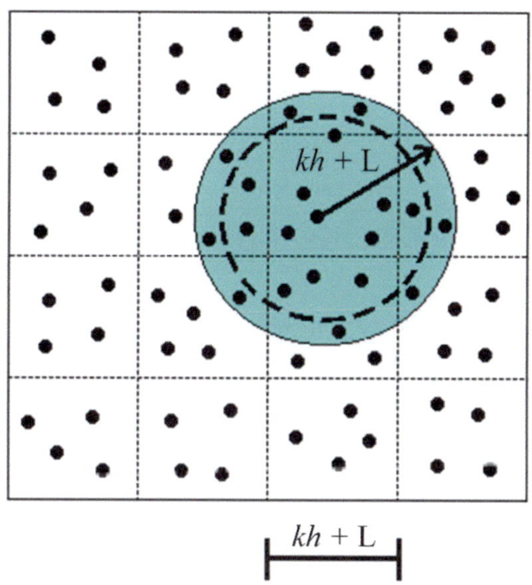

Fig. 3.5 The cut-off radius used in the Verlet list $(kh + L)$ and the probable neighbour particles storaged in the neighbour list

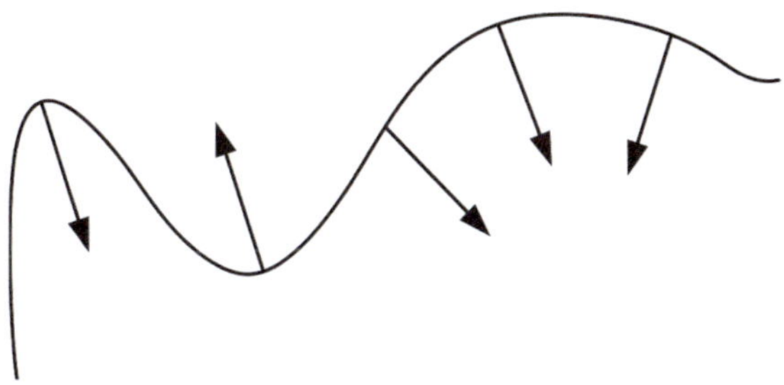

Fig. 3.6 Directions of force of surface tension at points with varying curvature

The method requires an additional quantity known as a colour field which has the value 1 at the particle locations and 0 at any other point in the space (outside the fluid), to find the vector normal to the surface.

Applying the SPH general expression (Eq. 3.15) to the colour field, we have the smoothed colour field:

$$Cs_i = \sum_{j=1}^{n} m_j \frac{Cs_j}{\rho_j} W(X_i - X_j, h) \tag{3.48}$$

where
Cs_i is the value of the colour field at the fixed particle
Cs_j is the value of the colour field at the neighbouring particle.

Inserting $Cs_j = 1$ into Eq. (3.48):

$$\boxed{Cs_i = \sum_{j=1}^{n} m_j \frac{1}{\rho_j} W(X_i - X_j, h)} \tag{3.49}$$

The gradient of the smoothed colour field yields a vector normal to the free surface at the position of the surface particle i, $\mathbf{n}_\mathrm{i}$:

$$\mathbf{n}_\mathrm{i} = \nabla Cs_i \tag{3.50}$$

The unit vector normal to the free surface at the position of the surface particle i, $(\mathbf{n}_\mathrm{u})_i$, is:

$$(\mathbf{n}_\mathrm{u})_i = \frac{\mathbf{n}_\mathrm{i}}{|\mathbf{n}_\mathrm{i}|} \tag{3.51}$$

The curvature of the free surface, κ, is given by the Laplacian of the smoothed colour field:

$$\kappa = -\nabla.(\mathbf{n}_\mathrm{u})_i = -\frac{\nabla^2 Cs_i}{|\mathbf{n}_\mathrm{i}|} \tag{3.52}$$

where the minus sign is required to ensure positive curvature and convex surfaces.

Putting everything together, the force of surface tension is calculated as follows:

$$\boxed{\mathbf{fs}_i = -\sigma \nabla^2 Cs_i \frac{\mathbf{n}_\mathrm{i}}{|\mathbf{n}_\mathrm{i}|}} \tag{3.53}$$

The colour field Cs_i and its gradient are used to identify the surface particles and to compute surface normals. A criterion used to identify if a particle i is on free surface is [19]:

$$|\mathbf{n}_\mathrm{i}| > l \tag{3.54}$$

where l is a threshold parameter.

Evaluating $n_i/|n_i|$ in Eq. (3.53) at locations where n_i is small causes numerical problems.

We only evaluate the force if $\mathbf{n}_i$ exceeds a certain threshold [19]. Therefore, the surface tension is only calculated if $|\mathbf{n}_i|$ is larger than a certain value. Changing this value would change the thickness of the "surface layer" of particles that were affected by surface tension. Fossum [20] experimented different values and a value of 20 seemed to create a reasonably thick layer.

In turn, the Laplacian of the colour field evaluated in the particle i in the free surface is approximated by SPH method, as follows:

$$\nabla^2 Cs_i = 2\sum_{j=1}^{n} \frac{m_j}{\rho_j}\left[Cs(X_i) - Cs\left(X_j\right)\right]\frac{\left(X_i - X_j\right)}{\left|X_i - X_j\right|^2}.\nabla W(X_i - X_j, h) \quad (3.55)$$

A new term related to the force of surface tension must be added to the momentum conservation equation of the particles on the free surface. Thus, the new equation to be solved is Eq. (3.56).

$$\boxed{\begin{aligned} \frac{D\vec{v}_i}{Dt} &= \sum_{j=1}^{n} m_j\left[\frac{P_i}{(\rho_i)^2} + \frac{P_j}{\left(\rho_j\right)^2}\right]\nabla W(X_i - X_j, h) + \\ 2\upsilon_i \sum_{j=1}^{n} \frac{m_j}{\rho_j}&\left(\vec{v}_i - \vec{v}_j\right)\frac{\left(X_i - X_j\right)}{\left|X_i - X_j\right|^2}.\nabla W(X_i - X_j, h) + \vec{g} + \mathbf{fs}_i \end{aligned}} \quad (3.56)$$

Morris [21] and Zhang et al. [22] implemented interfacial models, performed computational tests and provided results of surface tension simulations using SPH.

3.2.8.2 Incompressible Smoothed Particle Hydrodynamics (I-SPH) Method

A second methodology for the treatment of free surface has been presented by Ataie-Ashtiani et al. [23]. The researchers used an incompressible Smoothed Particle Hydrodynamics (I-SPH) formulation to simulate free surface incompressible fluid problems. The equations of conservation of mass and momentum were solved using a two-step fractional method. In the first step, the velocity field was computed without enforcing incompressibility. In the second step, a Poisson equation of pressure has been used to satisfy the incompressibility condition. The source term in the Poisson equation for the pressure was approximated, based on the SPH continuity equation, by an interpolation summation involving the relative velocities between a reference particle and its neighbouring particles. For a better stability of the simulations, a new form of the source term for the Poisson equation has been proposed and a modification in the Poisson equation of pressure has been performed in order to satisfy the incompressibility condition of the free surface particles.

The particles on the free surface do not have a complete domain of influence. Due to this, the particle density decreases near this boundary. A value chosen for a free surface parameter defines a criterion for the identification of the particles that make up that interface:

$$\boxed{\rho_* < \beta \rho_0} \tag{3.57}$$

where
β is the free surface parameter
ρ_* and ρ_0 and are the densities of the particle in the current and previous iteration, respectively.

The parameter β ranged from 0.80 to 0.99 in the simulations.

In [23], the I-SPH scheme has been used to simulate a 2-D dam breaking, the evolution of an elliptical water bubble, a solitary wave breaking on a mild slope and a tsunami run-up. The results obtained were considered satisfactory.

Lee et al. [24] applied the I-SPH model to identify the free surface particles in a moving particle semi-implicit method (MPS) implementation. In addition, a collision model based on the conservation of linear momentum was used in order to correctly identify the particles of the free surface in the flow over time. This collision model, in combination with other methods (particle shifting technique, kernel gradient correction method (see 3.4.1), free surface damping and artificial viscosity damping layer), has been implemented by [25] in SPH aiming to identify the free surface particles. This latter references present a complete analysis of the results found.

3.2.8.3 Surface Particle Tracking Using Divergence of the Positions of the Particles

Another technique presented in the literature ([26–28]) identifies the particles on the free surface from the value of SPH divergence of the positions of the particles, as shown in Eq. (3.58)

$$\boxed{\nabla.\mathbf{r}_i = \sum_{j=1}^{n} \frac{m_j}{\rho_j} (\mathbf{r}_i - \mathbf{r}_j).\nabla W(\mathbf{r}_i - \mathbf{r}_j, h)} \tag{3.58}$$

where
$\nabla.\mathbf{r}_i$ is the SPH divergence of the position of the fixed particle
$\mathbf{r}_i$ is the position of the fixed particle
$\mathbf{r}_j$ is the position of the neighbouring particle.

In 2-D domains, this divergence is equal to 2.0 in the core of the domain. Near the water surface, the number of neighbouring particles is insufficient (due to kernel truncation) and the divergence $\nabla.\mathbf{r}_i$ becomes lower than 2.0. A threshold value ranging from 1.5 to 1.2 is used to determine which particles belong to the surface. In identifying the free surface particles with this method, most free surface particles are captured correctly but not all of them. This defect is intrinsic as the undetected free surface particles still have a pressure very close to zero.

Most surface particles on the surface are identified but not all of them; this defect is acceptable, as the undetected surface particles still have a pressure very close to zero [26].

After the free surface particles identification is completed, the zero constant pressure condition can be imposed on those particles.

3.2.9 Treatment of the Interfaces

For modelling systems with multiple materials, an efficient algorithm is needed to describe the dynamic interfaces of the material. SPH particles are interpolation points in which the mass of each particle is agglomerated. A Lagrangian particle has a finite mass and density and, thus, a finite volume.

A particle has no definite shape. On the other hand, according to the interface treatment technique presented by [4], the shape of a particle may also be associated with a domain of influence, defined by the support radius (kh) of the Lagrangian element. Generally, a particle is considered to be circular (or spherical). This approach was used in this work.

The mass of a particle is usually considered as a constant. The updating of the density signifies the change of volume and size of the particle. Different shapes and sizes of particles increase the difficulty of the implementation of the contact algorithm. In order to make it simpler, only models with circular or spherical shaped particles are discussed here. SPH particles are considered to be rigid spheres rather than soft spheres, since soft sphere models also increase the difficulty of implementing the contact algorithm.

In the traditional SPH method, if the particles of different materials are not bonded together, the particles of one material can influence and can also be influenced by the particles of the other material.

Figure 3.7 shows an interface example of a material at the initial stage and at a later stage of evolution. Initially, the SPH particles are distributed regularly and penetration and deformation of the Lagrangian elements are not seen.

A contact algorithm includes the following steps: boundary identification, contact detection and response to the contact.

The detection of the particle–particle contact is performed by comparing the distance between the centres of mass of two particles at the interface to the sum of the radii that define the domains of influence. As shown in Fig. 3.8, the contact and penetration between two particles of fluids A and B can be identified by:

$$\mathrm{R} = \frac{(kh)_a + (kh)_b}{r_{ab}} \geq 1.0 \tag{3.59}$$

where
R is a parameter used in the detection of the particle–particle contact
r_{ab} is the distance between the centres of mass of two fluid particles at the interface.

A restorative contact force $\mathbf{f}_{ab}$ has been employed in many simulations as an analogy to the Lennard-Jones molecular force. It is applied along the centre line of the two particles in the event of contact or penetration. Liu et al. [29] used this force to simulate underwater explosion when it comes to different material interfaces.

$$\mathbf{f}_{ab} = \begin{cases} \bar{P}\left(\mathrm{R}^{N_1} - \mathrm{R}^{N_2}\right)\dfrac{X_{ab}}{{r_{ab}}^2}, & \text{if } \mathrm{R} \geq 1.0 \\ 0, & \text{in the other case.} \end{cases} \tag{3.60}$$

where
X_{ab} is the difference between the position vectors of the particles of fluids A and B
$\bar{P}$, N_1 and N_2 are commonly considered as 1×10^5, 6 and 4, respectively.

In the literature, there are various studies employing the repulsive forces, analogous to intermolecular forces, in the treatment of interfaces (see, for example, [30,31]). This treatment involves again (as was the case with certain boundary treatment techniques) an application of concepts of molecular modelling in continuum mechanics. Particles defined in the continuum domain exert forces analogous to

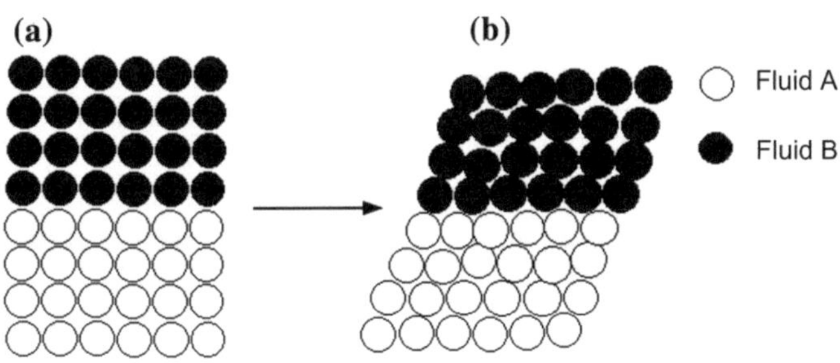

Fig. 3.7 Illustration of the material interface. **a** Initial distribution and **b** evolution of the particles

Fig. 3.8 Illustration of the contact and penetration between particles

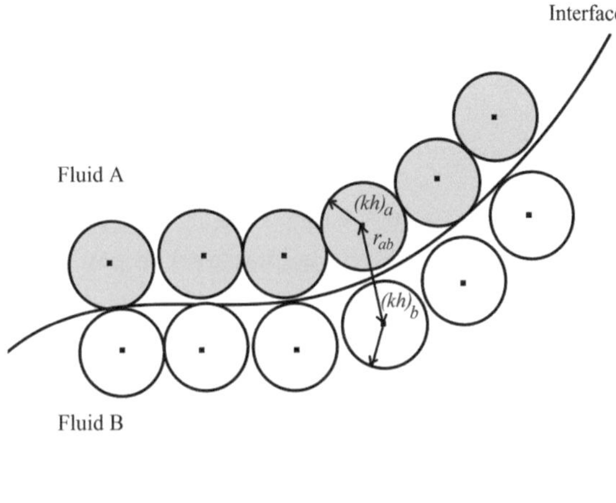

intermolecular forces on the microscopic scale. This represents a mixture of concepts from two different scales (microscopic and macroscopic) that has been applied for solving problems in continuum mechanics with the SPH method.

Reference [4] presents some other techniques used in the treatment of the interfaces in SPH.

3.2.10 Turbulence

Studies in turbulence are still new in SPH. Attempts to develop modelling for the turbulence in two and three dimensions have been made in recent years.

The direct numerical simulation (DNS) can be used in problems involving SPH method without any turbulence modelling. In other words, no modelling is required besides the physical conservation equations. This means that the conservation equations are solved in all the spatial points defined in the domain, for all temporal and spatial scales of turbulence.

However, the resolution of a wide range of scales of time and space time requires an extremely refined spatial and temporal discretization and a great computational effort. Turbulent flows have high degrees of freedom, which correspond to the number of linear equations discretized to be solved at each spatial point for a good flow characterization. Due to these high degrees of freedom, DNS leads to good results for low Reynolds numbers, which is a small fraction of the problems in the universe of fluid dynamics.

The study of an incompressible fluid flow inside a square cavity (lid-driven cavity problem) shows the good applicability of the direct numerical simulation in SPH, for low Reynolds numbers [32].

One of the oldest studies was conducted by Monaghan [33] for modelling turbulence in a compressible flow. In this study, an SPH version of the alpha turbulence model presented by Chen et al. [34] has been formulated for compressible flow with a resolution that varies in space and time. The alpha model involves two velocity fields: the first is obtained from the momentum equation and the other by the average of this velocity field as in the version of SPH called XSPH. The particles (fluid elements) are moved at the average velocity. The system satisfies a discrete Kelvin circulation theorem identical to that obtained with no velocity averaging. Despite providing good results, this method is not very computationally efficient [35].

Dalrymple and Rogers [36] presented a new method in which the standard SPH viscous formulation was replaced by the introduction of a subparticle scaling (SPS) technique using the large eddy simulation (LES) approach. Two-dimensional simulations of waves overtopping and waves breaking on the beach and three-dimensional simulations of dam breaking over structure and waves on a beach have been performed. The method presented good results in relatively small regions, where the particles can be held in a reasonable number. It is not recommended for large areas, which are more efficiently modelled by extended Boussinesq codes for example.

Violeau and Issa [37] made a review of turbulence models applied to SPH until that year. They presented three different models: the SPH $\kappa - \varepsilon$ model, the explicit algebraic Reynolds stress model (EARSM) and large eddy simulation (LES) model.

Despite its simplicity, the results provided by the SPH $\kappa - \varepsilon$ model showed a good agreement with the experiments in a simulation of a turbulent 2-D steady periodic open-channel flow. The collapse of a water column was also simulated using this model, which was validated qualitatively (comparison to experiments revealed a satisfactory agreement). In the end of the collapse of a water column simulation, it was concluded that SPH $\kappa - \varepsilon$ can treat complex flows involving highly disturbed free surfaces and rapid motion. However, no validation can be done in terms of shape of the free surface, which is of a great interest for all possible applications in environmental fluid mechanics prediction of wave run-up and overtopping, for example).

A more accurate approach, still based on a Reynolds-averaging idea, the explicit algebraic Reynolds stress model (EARSM) has been employed in simulations. In the collapse of a water column, the results obtained by EARSM showed that increasing the turbulence closure complexity improved significantly the accuracy of the free surface prediction, as required. However, due to the fact that the location of the free surface was performed from experimental images, the accuracy of the model could not be truly evaluated. The authors clarified that it is necessary to perform a more accurate and systematic model validation.

The main idea of the large eddy simulation (LES) is to simulate the large scales of turbulent motions while modelling the dissipative effect of smaller ones. This is justified by the fact that large scales carry the main part of turbulent kinetic energy and are most likely anisotropic, whereas small scales are believed to be almost isotropic. It provides much more detailed information on the turbulence itself and is more suitable for far-from-equilibrium flows, while considering reasonable computational costs (whereas in the particular of the SPH method, the cost can be high). The model was also applied to a 3-D collapse of water column. The results of the simulation showed the evolution of the free surface, revealing the realistic breaking wave when compared to the experimental data. The three-dimensional SPH approach together with the LES model led to a high computational cost, due to the numerous filtering operations. The authors pointed out the need to use parallel algorithms to enable later work using this model.

Monaghan [38] presented the SPH-ε model based on a Lagrangian similar to that used for the Lagrangian averaged Navier–Stokes (LANS) turbulence model but with a different smoothed velocity. The smoothed velocity preserves the shape of the spectrum of the unsmoothed velocity, but reduces the magnitude for short length scales by an amount which depends on a parameter ε. The results of SPH-ε model reproduced the results obtained by other researchers for decaying turbulence in a square box with non-slip boundary conditions. The turbulence model not only reproduces these results but, in addition, shows that quantities such as the second-order velocity correlation function calculated with an initial particle spacing of 1/150 can be reproduced with an initial particle spacing 1/75 which requires a factor 8 less computation to integrate to a specified time.

Direct numerical simulation of decaying turbulence in a non-slip square box has been performed by Robinson and Monaghan [39]. The results of SPH simulations were compared to published results using a high-accuracy pseudo-spectral code.

Ensemble averages of the kinetic energy, enstrophy and average vortex wavenumber compared well against the pseudo-spectral results, as did the evolution of the total angular momentum of the fluid.

In the absence of viscosity, standard SPH simulations produce purely noisy particle motion and at finite viscosities the method overpredicts dissipation. Aiming to solve this, Adami et al. [40] introduced a modified transport velocity to advect particles that homogenizes the particle distribution, thus stabilizing the numerical scheme. The authors performed two- and three-dimensional simulations and presented results of the Taylor–Green vortex flow. The SPH results for decaying turbulence, energy spectra and dissipation rates showed a good agreement with DNS data provided by the literature.

Mayrhofer et al. [41] performed simulations of three-dimensional turbulent channel flows using two different turbulence models. They were: a quasi-direct numerical simulation (DNS) of a channel with reduced size (whose results reproduced the statistical turbulence correctly except for some fluctuations in the near wall region) and a large eddy simulation (LES) of a standard size channel. In this second turbulence approach, the statistical turbulence was not correctly reproduced. Then, the Taylor–Green vortex flow was used and the authors concluded that the inexact prediction for turbulence may have been caused by lack of spatial resolution. However, the increase in this in order to reduce the error in the velocity−pressure−gradient correlation terms would lead to a very high number of fluid particles, in contrast to the expected moderate resolution for three-dimensional SPH applications.

3.2.11 Variable Smoothing Length

The length of the domain of influence is very important in the SPH method. It influences the efficiency of the calculations and the accuracy of the solutions. If the length of the domain of influence is too small or too large, the results will not be consistent with the physical problem studied.

In the simulation of problems where there are shocks or impacts with sudden local variation of density, it is necessary to vary the smoothing length, so that the number of neighbouring particles is maintained around a constant value.

In the literature, many ways to vary dynamically the smoothing length, aiming that the number of the neighbouring particles remains relatively constant, are proposed [3]. The simplest approach is to update h according to the averaged density:

$$\boxed{h = h_0\left(\frac{\rho_{zero}}{\rho}\right)^{\frac{1}{n_d}}} \tag{3.61}$$

where

h_0 is the initial smoothing length

ρ_{zero} is the initial density

n_d is the number of dimensions of the domain.

Another method takes the time derivative of the smoothing length (h) in terms of the continuity equation:

$$\boxed{\frac{Dh}{Dt} = -\frac{1}{n_d}\frac{h}{\rho}\frac{D\rho}{Dt}} \tag{3.62}$$

3.2.12 Numerical Aspects and Corrections

In SPH simulations, certain methods for corrections and ensurance of convergency can be employed. In the literature, there are many tools employed and results available.

There is no rule used in the definition of numerical corrections to be employed in the simulations. In fact, there are solutions for specific problems in which the numerical corrections employed are listed. In the next subsections, some numerical corrections applied in SPH simulations and references for study are briefly presented. However, it is important to alert the reader that the use of artificial tools for corrections of the numerical physical-mathematical model implemented is liable to criticism.

The justifications for the use of these numerical and artificial corrections should be provided based on the inaccuracy or non-convergence of the numerical results provided by the implemented SPH scheme. It is recommended that this analysis and justification be carried out with prudence and understanding of the physical, mathematical and numerical modelling, for each studied problem.

3.2.12.1 Artificial Viscosity

The transformation of kinetic energy into heat takes place in problems involving mainly shock waves and needs to be measured, which does not happen when Eqs. (3.38) and (3.39) are employed.

That energy transformation can be represented as a form of viscous dissipation. The use of the artificial viscosity application in the simulations aims to avoid numerical instabilities and the interpenetration between particles. The formulation used in the modelling of artificial viscosity is shown in Eq. (3.63) [3].

$$\pi_{ij} = \begin{cases} \dfrac{-\alpha_\pi c_{ij}\chi_{ij} + \beta_\pi \chi_{ij}^2}{\bar{\rho}_{ij}}, \ \vec{\mathrm{v}}_{ij}.X_{ij} < 0 \\ 0, \ \vec{\mathrm{v}}_{ij}.X_{ij} > 0 \end{cases} \tag{3.63}$$

$$\chi_{ij} = \frac{h_{ij}\vec{\mathrm{v}}_{ij}.X_{ij}}{\left|X_{ij}\right|^2 + \varphi^2} \tag{3.64}$$

$$c_{ij} = \frac{c_i + c_j}{2} \tag{3.65}$$

$$\bar{\rho}_{ij} = \frac{\rho_i + \rho_j}{2} \tag{3.66}$$

$$h_{ij} = \frac{h_i + h_j}{2} \tag{3.67}$$

$$\vec{\mathrm{v}}_{ij} = \vec{\mathrm{v}}_i - \vec{\mathrm{v}}_j \tag{3.68}$$

where
π_{ij} is the artificial viscosity
α_π and β_π are coefficients used in the calculation of artificial viscosity
c_i e c_j are the velocities of sound in the fixed and neighbouring particles, respectively
h_i e h_j are the support radii of the fixed and neighbouring particles, respectively
$\vec{\mathrm{v}}_{ij}$ is the relative velocity between the particles i and j
φ^2 is a factor that prevents numerical differences when two particles approach each another.

In the previous equations, α_π e β_π are constants whose values vary according to the problem studied and the factor φ^2 is set to $0.01{h_{ij}}^2$.

The term related to artificial viscosity is added to the terms of pressure in the SPH approximations to the momentum and energy conservation equations, Eqs. (3.38) and (3.39). After its addition to the momentum conservation equation, we obtain the following equation for the acceleration of the particle:

$$\boxed{\frac{\mathrm{D}\vec{\mathrm{v}}_i^{\,*}}{Dt} = \frac{\mathrm{D}\vec{\mathrm{v}}_i}{Dt} - \sum_{j=1}^{n} m_j \pi_{ij} \nabla W(X_i - X_j, h)} \tag{3.69}$$

where $\frac{\mathrm{D}\vec{\mathrm{v}}_i^{\,*}}{Dt}$ is the acceleration of the fixed particle, after the addition of the term related to the artificial viscosity to the momentum conservation equation.

The first term on the right-hand side of Eq. (3.69) is calculated from Eq. (3.38).

Analogously, the artificial viscosity term is added to the energy conservation equation:

$$\boxed{\frac{\mathrm{De}_i^{\,*}}{Dt} = \frac{\mathrm{De}_i}{Dt} + \sum_{j=1}^{n} m_j \pi_{ij} \left(\vec{\mathrm{v}}_i - \vec{\mathrm{v}}_j\right) . \nabla W(X_i - X_j, h)} \tag{3.70}$$

where $\frac{\mathrm{De}_i^{\,*}}{Dt}$ is the rate of change of specific internal energy of the fixed particle in time after the addition of the term related to the artificial viscosity to the energy conservation equation.

The first term on the right-hand side of the previous equation is calculated using Eq. (3.39).

The study of [42] presents the calibration of the α_π parameter (when using β_π = 0.00) in simulations of regular waves from the comparison with experimental data.

3.2.12.2 XSPH Correction

To avoid the disordered movement of the particles and to guarantee the exact conservation of the linear and angular momentum, Monaghan [43] proposed a corrected form for the velocities and, consequently, for the positions of the particles. This correction is known as XSPH. The particles are moved in a more ordered manner at the average velocity of their interaction partners, as shown in Eq. (3.71):

$$\boxed{\vec{v}_i^{\,*} = \vec{v}_i + \eta \sum_{j=1}^{n} \frac{2m_j}{\rho_i + \rho_j} \left(\vec{v}_j - \vec{v}_i\right) W(X_i - X_j, h)} \tag{3.71}$$

where
$\vec{v}_i^{\,*}$ is the velocity of the fixed particle, after the application of the XSPH correction
η is a parameter ranging from 0 to 1.0, being commonly used as 0.50.

In Paiva et al. [44], the effect of the XSPH correction application in a lava flow simulation can be clearly seen.

3.2.12.3 Tensile Instability

Tensile instability results in the formation of particle agglomerates, in an unphysical behaviour. It occurs in regions where the particles present negative pressures, coming from the prediction of pressure by the equation of state.

In the numerical simulations, the agglomerates formed have, in general, the form of rows of particles. The formation of anisotropic strings in the flow around a square cylinder can be seen in [45].

After performing simulations whose results showed good agreement with the experiments, Monaghan [46] proposed the addition of a new term to the momentum conservation equation, similar to an artificial pressure, in order to promote the repulsion between the particles and to avoid tensile instability. The term to be added to Eq. (3.38) is:

$$\boxed{-\sum_{j=1}^{n} m_j R_p (f_{ij})^s \nabla W(X_i - X_j, h)} \tag{3.72}$$

$$f_{ij} = \frac{W(X_i - X_j, h)}{W(\Delta p, h)} \tag{3.73}$$

where
Δp is the average spacing in the neighbourhood of the fixed particle
s assumes the value 4, for simulations in fluid dynamics [46]

$$R_p = R_{p_i} + R_{p\,j} \tag{3.74}$$

$$R_{p_i} = \begin{cases} \lambda \dfrac{|P_i|}{\rho_i^2}, & \text{se } P_i < 0 \\ 0, & \text{in the other case.} \end{cases} \tag{3.75}$$

$$R_{p_j} = \begin{cases} \lambda \dfrac{|P_j|}{\rho_j^2}, & \text{se } P_j < 0 \\ 0, & \text{in the other case.} \end{cases} \tag{3.76}$$

where
P_i and P_j are the pressures acting on the fixed and neighbouring particle, respectively
λ is a parameter which value depends on h, s and n_p.

Lobovský and Kren [47] applied this numerical correction, which they called artificial stress, in hydrodynamics (Couette, Poiseuille and free surface flows) and elastic collisions between rubber rings (modelled as elastic materials).

Swegle [48] presented an analysis of the tensile instability and a comparison of the occurrence of this phenomenon in SPH and moving least square methods.

Instabilities in the stress states similar to the tensile instability found in the SPH simulations were seen by Khayyer and Gotoh [49] solving hydrodynamics problems using another well-known particle method: the moving particle semi-implicit (MPS) method.

Chen et al. [50] proposed the utilization of the CSPM method (see 3.4.1.1) as a way to mitigate the tensile instability.

Yang et al. [51] employed a hyperbolic-shaped kernel aiming to remove the undesirable phenomenon.

3.3 Temporal Integration Methods

The updating of the properties of the particles in time (position, velocity, energy and others) was made by temporal numerical integration methods. Among the methods of temporal integration that can be employed, there are Runge–Kutta of first order (Euler), leap frog and more elaborate algorithms such as the predictor–corrector, Verlet, symplectic and Beeman schemes.

Next, the integration schemes of the Euler, leapfrog and predictor–corrector methods will be presented. The description of Verlet, symplectic and Beeman schemes can be found in [52].

3.3.1 Euler's Integration Method

In this method, also called Runge–Kutta of first order, the velocity and position are updated in parallel, as follows:

$$\begin{cases} \vec{\mathrm{v}}(t+\Delta t) = \vec{\mathrm{v}}(t) + \dfrac{D\vec{\mathrm{v}}}{Dt}(\Delta t) \\ \\ X(t+\Delta t) = X(t) + \Delta t\,\vec{\mathrm{v}}(t+\Delta t) \end{cases} \tag{3.77}$$

where Δt is the integration time step

As a negative point, Euler's method presents first-order accuracy and can become quite unstable under certain circumstances, which can be corrected by taking a very small time step Δt. The stability of the integration depends on the choice of the integration time step Δt. The CFL numerical stability criterion (Courant–Friedrichs–Lewy condition) is applied to guarantee the convergence of results [53].

Different criteria are used in the choice of the time step in SPH simulations. One of them requires the time step to be proportional to the smallest spatial particle resolution, which in SPH applications is represented by the smallest smoothing length ($\mathrm{h_i}$) [3].

$$\Delta t = \min\left(\frac{\mathrm{h_i}}{c}\right) \tag{3.78}$$

An expression to estimate the time step based on viscous diffusion is presented in [8]:

$$\Delta t = 0.125\left(\frac{h^2}{\nu}\right) \tag{3.79}$$

Monaghan and Koss [54] proposed a variable time step controlled by a Courant condition and a viscosity condition. The reference [52] also employed a variable time step in simulations.

Other proposals of the time step for simulations can be found in [3,5].

3.3.2 Leapfrog Method

The velocity is calculated at the midpoints of the time intervals. This way, position and velocity are evaluated alternately.

$$\begin{cases} \vec{v}(t+\frac{1}{2}\Delta t) = \vec{v}(t-\frac{1}{2}\Delta t) + \frac{D\vec{v}}{Dt}(\Delta t) \\ X(t+\Delta t) = X(t) + \Delta t\,\vec{v}(t+\frac{1}{2}\Delta t) \end{cases} \tag{3.80}$$

The velocity is calculated as the average between the velocities at the previous and posterior instants of time:

$$\vec{v}(t) = \frac{1}{2}\left[\vec{v}(t+\frac{1}{2}\Delta t) + \vec{v}(t-\frac{1}{2}\Delta t)\right] \tag{3.81}$$

Initializing the method requires an initial velocity calculation $\vec{v}_i(-1/2)$ by the Euler method:

$$\vec{v}(-1/2) = \vec{v}(0) - \frac{1}{2}\frac{D\vec{v}}{Dt}\bigg|_{t=0}(\Delta t) \tag{3.82}$$

3.3.3 Predictor–Corrector

The predictor–corrector or improved Euler's method uses an explicit method for the predictor step and an implicit method for the corrector step. It does not advance a time step directly.

At the beginning, physical quantities must be known at the current time instant (t_o). Then, the physical properties are predicted at the intermediate instant $\left(t_o + \frac{\Delta t}{2}\right)$ using the Euler scheme:

$$\begin{cases} X_i{}^P\left(t_o+\frac{\Delta t}{2}\right) = X_i(t_o) + \frac{\Delta t}{2}\left(\frac{DX_i}{Dt}\right)_{t_o} \\ \vec{v}_i{}^P\left(t_o+\frac{\Delta t}{2}\right) = \vec{v}_i(t_o) + \frac{\Delta t}{2}\left(\frac{D\vec{v}_i}{Dt}\right)_{t_o} \\ \rho_i{}^P\left(t_o+\frac{\Delta t}{2}\right) = \rho_i(t_o) + \frac{\Delta t}{2}\left(\frac{D}{\rho_i}Dt\right)_{t_o} \\ e_i{}^P\left(t_o+\frac{\Delta t}{2}\right) = X_i(t_o) + \frac{\Delta t}{2}\left(\frac{De_i}{Dt}\right)_{t_o} \end{cases} \tag{3.83}$$

where $\vec{\mathrm{v}}_i^{\,P}$, ${\rho_i}^P$, ${X_i}^P$ and e_i^P are the predictions of the velocity, density, position and specific internal energy of the fixed particle, respectively.

After the first step, it is necessary to calculate $P_i\left(t_o+\frac{\Delta t}{2}\right)$ as a function of $\left(\rho_i^P\left(t_o+\frac{\Delta t}{2}\right)\right)$, employing the Tait equation.

The quantities $\frac{\mathrm{D}\vec{\mathrm{v}}_i}{Dt}$, $\frac{\mathrm{D}\rho_i}{Dt}$, $\frac{\mathrm{D}X_i}{Dt}$ and $\frac{\mathrm{D}e_i}{Dt}$ must be obtained at the intermediate instant $\left(t_o+\frac{\Delta t}{2}\right)$.

Finally, the physical quantities are obtained at the final time instant $(t_o+\Delta t)$ as follows:

$$\begin{cases} X_i\,(t_o+\Delta t)=X_i(t_o)+\dfrac{\Delta t}{2}\left[\left(\dfrac{\mathrm{D}X_i}{Dt}\right)_{t_o}+\left(\dfrac{\mathrm{D}X_i}{Dt}\right)_{(t_o+\frac{1}{2}\Delta t)}\right] \\ \vec{\mathrm{v}}_i\,(t_o+\Delta t)=\vec{\mathrm{v}}_i(t_o)+\dfrac{\Delta t}{2}\left[\left(\dfrac{\mathrm{D}\vec{\mathrm{v}}_i}{Dt}\right)_{t_o}+\left(\dfrac{\mathrm{D}\vec{\mathrm{v}}_i}{Dt}\right)_{(t_o+\frac{1}{2}\Delta t)}\right] \\ \rho_i\,(t_o+\Delta t)=\rho_i(t_o)+\dfrac{\Delta t}{2}\left[\left(\dfrac{\mathrm{D}\rho_i}{Dt}\right)_{t_o}+\left(\dfrac{\mathrm{D}}{\rho}_i\,Dt\right)_{(t_o+\frac{1}{2}\Delta t)}\right] \\ e_i\,(t_o+\Delta t)=e_i(t_o)+\dfrac{\Delta t}{2}\left[\left(\dfrac{De_i}{Dt}\right)_{t_o}+\left(\dfrac{De_i}{Dt}\right)_{(t_o+\frac{1}{2}\Delta t)}\right] \end{cases} \tag{3.84}$$

3.4 Consistency

According to the equivalence theorem of Lax-Richtmyer, it is known that if a numerical model is stable, the convergence of the solution to the well-posed problem will be determined by the consistency of the approximation function. In the SPH method, besides depending on the approximation function employed, the consistency also depends on the influence domain, the number of particles and the distribution within it.

Analysing the approximation of a function by the SPH method, obtained by the expansion of the Taylor series, Eq. (3.14), we obtain a set of conditions for the kernel moments up to the Nth order (M_N) [3].

The consistencies of the zeroth and first orders (C^0 and C^1, respectively) are guaranteed to the kernel if the values of the moments of order 0 and 1 (M_0 and M_1) are those presented in Eq. (3.85).

$$\begin{cases} M_0 = \int\limits_{\Omega} W(X - X', h)dX' = 1, \\ M_1 = \int\limits_{\Omega} (X - X')W(X - X', h)dX' = 0, \\ M_2 = \int\limits_{\Omega} (X - X')^2 W(X - X', h)dX' = 0, \\ \quad \vdots \\ M_N = \int\limits_{\Omega} (X - X')^2 W(X - X', h)dX' = 0. \end{cases} \tag{3.85}$$

If the domain of influence is continuous and complete, there is a uniform distribution of particles inside it, and the support radius has been properly defined, and the conditions for the consistencies of zeroth and first orders are met due to the properties of normalization and symmetry of the kernel. However, when the approximations are extended to regions truncated on the contours, not even the consistency of order 0 is guaranteed. In addition, satisfying the kernel consistency conditions does not necessarily mean that the consistency of the SPH method is valid.

Figure 3.9 shows how the truncation of the domain of influence occurs. In boundary regions, as shown in Figure 3.9b, the consistency of the kernel is not guaranteed. In addition, there is also a number of particles that does not occupy the entire domain of influence. This phenomenon is known as particle inconsistency and leads to less accurate approximations of physical properties.

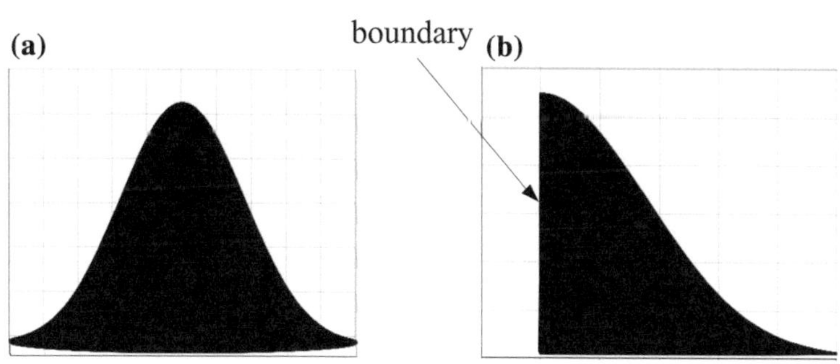

Fig. 3.9 Domain of influence in an two-dimensional case: **a** complete and **b** truncated

The discrete counterparts of the constant and linear consistencies (zeroth and first orders) conditions are presented in:

$$\boxed{\sum_{j=1}^{n} W(X_i - X_j, h)\Delta V_j = 1} \tag{3.86}$$

$$\boxed{\sum_{j=1}^{n} (X_i - X_j)W(X_i - X_j, h)\Delta V_j = 0} \tag{3.87}$$

where ΔV_j is the volume of each Lagrangian neighbouring particle.

These two consistency conditions are generally not satisfied. Respecting the minimum number of neighbouring particles within the domain of influence (commented in Sect. 3.2), when there is an imbalance of particles within the domain of influence (with the existence of more dense regions), the left-hand side of Eq. (3.86) is less than 1 and the left-hand side of Eq. (3.87) is not null. The same occurs when there is truncation of the domain of influence, a situation shown in Fig. 3.9b.

The use of the appropriate support radius influences the consistency of the SPH method. The influence of the support radius in the approximations obtained by SPH needs to be properly analysed [4].

3.4.1 Restoration of the Consistency

In the fluid simulations using SPH, there are a finite spacial domain and free surfaces. The fluid particles near the boundaries or the free surfaces have a kernel smoothing function truncated due to the absence of neighbouring particles. Some methods employed for the restoration of the consistency are presented below.

3.4.1.1 Corrective Smoothed Particle Method

A proposal to restorate the consistency in the SPH method was presented by Chen et al. [55]. The method was called Corrective Smoothed Particle Method (CSPM) and had as its starting point the expansion of the Taylor series around the fixed position X.

$$f(X') = f(X) + f'(X)(X' - X) + \frac{f''(X)}{2!}(X' - X)^2 + \cdots \tag{3.88}$$

where

$f'(X)$ and $f''(X)$ are the first- and second-order derivatives of the function evaluated in the position X, respectively

Multiplying both sides by the kernel and integrating over the domain:

$$\int_{\Omega} f(X')W(X' - X, h)dX' = f(X)\int_{\Omega} W(X' - X, h)dX' +$$

$$f'(X)\int\limits_{\Omega}(X'-X)W(X'-X,h)dX'+\frac{f''(X)}{2!}\int\limits_{\Omega}(X'-X)^2W(X'-X,h)dX'+\cdots \tag{3.89}$$

For the inner regions of the domain, in which the kernel truncation does not occur, the second term on the right-hand side of Eq. (3.89) is null due to the symmetry property of the kernel. Near the boundaries, that term will be nonzero.

Disregarding all terms with derivatives of the Taylor expansion, an expression for the correction of $f(X)$ is obtained:

$$f(X)=\frac{\int_{\Omega}f(X')W(X'-X,h)dX'}{\int_{\Omega}W(X'-X,h)dX'} \tag{3.90}$$

Comparing Eqs. (3.14) and (3.90), it can be seen that, for the interior regions of the domain (without kernel truncation), the approximations of the function f provided by the SPH original (uncorrected) and by CSPM (corrected) are really the same, due to the satisfaction of the normalization condition. In the boundary regions, since the integral of the smoothing function is truncated by the contour, the normalization condition is not satisfied. With the implementation of the CSPM for the function correction, initially approximated by the traditional SPH, the zero-order consistency restoration is performed, given that the denominator of Eq. (3.90) is less than 1.0.

When replacing $W(X'-X,h)$, in Eq. (3.89), by $W'(X'-X,h)$ and neglecting the second-order and the higher derivatives, a correction for the first derivative of $f(X)$, in regions near the boundaries, is found:

$$f'(X)=\frac{\int_{\Omega}\left[f(X')-f(X)\right]W'(X'-X,h)dX'}{\int_{\Omega}(X'-X)W'(X'-X,h)dX'} \tag{3.91}$$

where $W'(X'-X,h)$ is the derivative of the kernel evaluated in position $(X'-X)$.

After the particle discretization, the expressions for the correction of the function f and its derivatives in the position occupied by the fixed particle, near the boundaries, are obtained:

$$\boxed{f_i=\frac{\sum_{j=1}^{n}f_jW(X_i-X_j,h)\dfrac{m_j}{\rho_j}}{\sum_{j=1}^{n}W(X_i-X_jh)\dfrac{m_j}{\rho_j}}} \tag{3.92}$$

$$\boxed{f'_i=\frac{\sum_{j=1}^{n}\left(f_j-f_i\right)W'(X_i-X_j,h)\dfrac{m_j}{\rho_j}}{\sum_{j=1}^{n}W'(X_i-X_j,h)(X_j-X_i)\dfrac{m_j}{\rho_j}}} \tag{3.93}$$

- **Density Renormalization**

In SPH, large pressure variations can be seen on the particles, which is particularly important near the boundaries and free surface.

A way to smooth pressure fluctuations is to filter the densities of the particles from the densities of the neighbours and to restart these properties after a number of iterations.

The particle density correction is made from Eq. (3.4.1.1), obtained using Eq. (3.92): The particle density correction is made from Eq. (3.92):

$$(\rho_i)_{filtered} = \frac{\sum_{j=1}^{n} \rho_j W(X_i - X_j, h)\frac{m_j}{\rho_j}}{\sum_{j=1}^{n} W(X_i - X_j, h)\frac{m_j}{\rho_j}} \Rightarrow$$

$$\boxed{(\rho_i)_{filtered} = \frac{\sum_{j=1}^{n} m_j W(X_i - X_j, h)}{\sum_{j=1}^{n} W(X_i - X_j, h)\frac{m_j}{\rho_j}}} \tag{3.94}$$

where $(\rho_i)_{filtered}$ is the density of the fixed particle after the density renormalization.

Analysing Eq. (3.94), we conclude that a renormalization of the densities of the particles, obtained from the solution of the mass conservation equation, is being carried out by CPSM method.

- **Correction of the Pressure Gradient**

Corrections of the pressure gradient components, in order to eliminate the effect of the contours and the kernel truncation, can be performed in a 2-D domain by the CSPM method using the following equations:

$$\boxed{\left(\frac{1}{\rho}\frac{\partial P}{\partial \mathrm{x}}\right)_i^* = \frac{\left(\frac{1}{\rho}\frac{\partial P}{\partial \mathrm{x}}\right)_i}{\sum_{j=1}^{n} \frac{\partial W(X_i - X_j, h)}{\partial \mathrm{x}}(\mathrm{x}_j - \mathrm{x}_i)\frac{m_j}{\rho_j}}} \tag{3.95}$$

$$\boxed{\left(\frac{1}{\rho}\frac{\partial P}{\partial \mathrm{y}}\right)_i^* = \frac{\left(\frac{1}{\rho}\frac{\partial P}{\partial \mathrm{y}}\right)_i}{\sum_{j=1}^{n} \frac{\partial W(X_i - X_j, h)}{\partial \mathrm{y}}(\mathrm{y}_j - \mathrm{y}_i)\frac{m_j}{\rho_j}}} \tag{3.96}$$

where

$\left(\frac{1}{\rho}\frac{\partial P}{\partial \mathrm{x}}\right)_i^*$ is the component of the pressure gradient in the fixed particle, corrected by the CSPM method, per unit mass, in the direction x

$\left(\frac{1}{\rho}\frac{\partial P}{\partial \mathrm{y}}\right)_i^*$ is the component of the pressure gradient in the fixed particle, corrected by the CSPM method, per unit mass, in the direction y

x_i is the abscissa of the fixed particle

x_j is the abscissa of the neighbouring particle

y_i is the ordinate of the fixed particle

y_j is the ordinate of the neighbouring particle.

3.4.1.2 Mean Least Squares (MLS)

This correction method was developed by Dilts [56] and applied successfully by Colagrossi and Landrini [57]. This is a first-order correction for the density fields [52]:

$$\boxed{(\rho_i)_{corrected} = \frac{\sum_{j=1}^{n} \rho_j W^{MLS}(X_i - X_j, h)\frac{m_j}{\rho_j}}{\sum_{j=1}^{n} W^{MLS}(X_i - X_j, h)\frac{m_j}{\rho_j}}} \tag{3.97}$$

where

$(\rho_i)_{corrected}$ is the density of the fixed particle after the MLS correction

$W^{MLS}(X_i - X_j, h)$ is the smoothing function provided by the mean least square (MLS) method evaluated at the position $(X_i - X_j, h)$

In a two-dimensional domain:

$$W^{MLS}\left(X_i - X_j, h\right) = \left[\beta_o(X_i) + \beta_{1\mathrm{x}}(X_i)\left(\mathrm{x}_i - \mathrm{x}_j\right) + \beta_{1\mathrm{y}}(X_i)\left(\mathrm{y}_i - \mathrm{y}_j\right)\right] W\left(X_i - X_j, h\right) \tag{3.98}$$

The correction vector $\beta(X_i)$ is given by:

$$\beta(X_i) = \begin{pmatrix} \beta_o(X_i) \\ \beta_{1\mathrm{x}}(X_i) \\ \beta_{1\mathrm{y}}(X_i) \end{pmatrix} = A^{-1}\begin{pmatrix} 1 \\ 0 \\ 0 \end{pmatrix} \tag{3.99}$$

$$A = \sum_{j=1}^{n} W\left(X_i - X_j, h\right)\tilde{A}\frac{m_j}{\rho_j} \tag{3.100}$$

$$\tilde{A} = \begin{pmatrix} 1 & \left(\mathrm{x}_i - \mathrm{x}_j\right) & \left(\mathrm{y}_i - \mathrm{y}_j\right) \\ \left(\mathrm{x}_i - \mathrm{x}_j\right) & \left(\mathrm{x}_i - \mathrm{x}_j\right)^2 & \left(\mathrm{y}_i - \mathrm{y}_j\right)\left(\mathrm{x}_i - \mathrm{x}_j\right) \\ \left(\mathrm{y}_i - \mathrm{y}_j\right) & \left(\mathrm{x}_i - \mathrm{x}_j\right)\left(\mathrm{y}_i - \mathrm{y}_j\right) & \left(\mathrm{y}_i - \mathrm{y}_j\right)^2 \end{pmatrix} \tag{3.101}$$

In a 3-D domain, the equations are similar, just including the z-direction.

3.4.1.3 Kernel Gradient Correction Method

In this method, the kernel gradient is corrected and after applied to the forces calculations in the equation of momentum conservation [52]:

$$\boxed{\tilde{\nabla} W_{ij} = L_i \nabla W_{ij}} \tag{3.102}$$

where
L_i is a matrix
$\tilde{\nabla} W_{ij}$ is the corrected kernel gradient

$$\boxed{L_i = M_i^{-1}} \tag{3.103}$$

$$W_{ij} = W\left(X_i - X_j, h\right) \tag{3.104}$$

$$M_i = \sum_{j=1}^{n} \frac{m_j}{\rho_j} \nabla W_{ij} \otimes \left(X_i - X_j\right) \tag{3.105}$$

In a 2-D domain, and using the polar coordinate system:

$$\nabla W_{ij} = \left. \frac{dW(r,h)}{dr} \right|_{r=r_{ij}} \frac{1}{r_{ij}} \left(X_i - X_j\right) \tag{3.106}$$

$$r_{ij} = \left|X_i - X_j\right| \tag{3.107}$$

where r is the radial direction

The components of the matrix M_i are:

$$M_i(1,1) = -\sum_{j=1}^{n} \frac{m_j}{\rho_j} \left. \frac{dW(r,h)}{dr} \right|_{r=r_{ij}} \frac{1}{r_{ij}} \left(x_i - x_j\right)^2 \tag{3.108}$$

$$M_i(1,2) = -\sum_{j=1}^{n} \frac{m_j}{\rho_j} \left. \frac{dW(r,h)}{dr} \right|_{r=r_{ij}} \frac{1}{r_{ij}} \left(x_i - x_j\right)\left(y_i - y_j\right) \tag{3.109}$$

$$M_i(2,1) = -\sum_{j=1}^{n} \frac{m_j}{\rho_j} \left. \frac{dW(r,h)}{dr} \right|_{r=r_{ij}} \frac{1}{r_{ij}} \left(x_i - x_j\right)\left(y_i - y_j\right) \tag{3.110}$$

$$M_i(2,2) = -\sum_{j=1}^{n} \frac{m_j}{\rho_j} \left. \frac{dW(r,h)}{dr} \right|_{r=r_{ij}} \frac{1}{r_{ij}} \left(y_i - y_j\right)^2 \tag{3.111}$$

The matrix M_i e and its inverse L_i are equal to the identity matrix when the fixed particle i is far from the boundaries or the free surface. In this case, the correction of the kernel gradient is not performed. Nevertheless, when the fixed particle is near the boundaries or the free surface, the distribution of particles around it is not symmetric anymore. Thus, both M_i and L_i are different from the identity matrix, and the kernel gradient is corrected as shows Eq. (3.112):

$$\begin{pmatrix} \tilde{\nabla} W_{ij}^{\mathrm{x}} \\ \tilde{\nabla} W_{ij}^{\mathrm{y}} \end{pmatrix} = \begin{pmatrix} L_i(1,1) & L_i(1,2) \\ L_i(2,1) & L_i(2,2) \end{pmatrix} \begin{pmatrix} \nabla W_{ij}^{\mathrm{x}} \\ \nabla W_{ij}^{\mathrm{y}} \end{pmatrix} \tag{3.112}$$

where
$\nabla W_{ij}^{\mathrm{x}}$ and $\nabla W_{ij}^{\mathrm{y}}$ are the components of the kernel gradient
$\tilde{\nabla} W_{ij}^{\mathrm{x}}$ and $\tilde{\nabla} W_{ij}^{\mathrm{y}}$ are the components of the corrected kernel gradient
x and y are subscripts that represent the coordinates of the 2-D domain.

3.5 Boundary Treatment Techniques

Appropriate boundary treatment technique is one of the greatest challenges found in the Smoothed Particle Hydrodynamics method. In this section, different boundary treatment techniques will be presented and commented.

Regarding the solid boundary treatments in SPH, they are often modelled using fictitious particles ([54,58–62]). Virtual, ghost, mirror, dummy and dynamic particles can be framed in this class of particles distributed over or near the contours which exert forces on the real particles (that discretized the mass of fluid) by keeping them within the domain. The forces that the created particles exert on the fluid particles are, in general, artificial, inspired by intermolecular forces models.

There is still a second approach, in which the dynamic particles obey the laws of conservation that govern the movement of real particles influencing their physical properties and forces acting on them, through the interpolations carried out in the particle method [63].

Besides these boundary treatment techniques, there are others in the literature. Currently, there is a search for new techniques that do not use fictitious particles in the boundary treatment in the most advanced research groups. Some techniques that deserve to be studied, in order to verify their concordance with the concepts of continuum mechanics, are: a semi-analytical model with the evaluation of the contact forces in rigid boundaries [64]; the unified semi-analytical boundary conditions, an enhanced form of the previous technique with wall-corrected gradients and possible to be used in a general boundary shape in 2-D [65,66] and 3-D [67]; a boundary integral formulation presented by [68] and a new SPH modelling proposed by [69] for viscous and non-viscous flows in the presence of 3-D complex boundaries.

The techniques using the fictitious, repulsive and dynamic particles, and the reflective treatment will be presented in the next subsections. A brief description of the open periodic boundary conditions also will be provided.

3.5.1 Fictitious Particles and Artificial Repulsive Forces

One way to carry out the boundary treatment of solid closed contours in SPH consists in the use of boundary (virtual/fictitious) particles on the contours or near them (in an extended region of the domain).

According to their positions, the virtual particles are classified into two categories: type I, located on the contour, and type II, located outside the boundary. Figure 3.10 shows both types of virtual particles in the simulation of solid walls.

The virtual particles of type II (or ghost particles) are not related to forces. They can be produced symmetrically according to the corresponding real particles in each evolution step. They have been used in SPH to treat solid boundaries and free surfaces [3] and [70]).

After the initial set-up of all particles in the domain, physical properties can be attributed to the virtual particles, for example velocity and temperature (Dirichlet boundary conditions) and pressure on the solid walls (Neumann boundary conditions) [26]. From the initial properties of the fictitious particles, approximate properties of the fluid particles can be obtained.

Depending on the problem studied, artificial repulsive forces have been employed in order to prevent the real particles from exceeding the boundaries and escaping the domain. In general, the repulsive forces are calculated by an analogy with the Lennard-Jones molecular force [3,71]). A virtual particle of type I, located on the boundary, exerts a force on the fluid particle in the direction of the line connecting

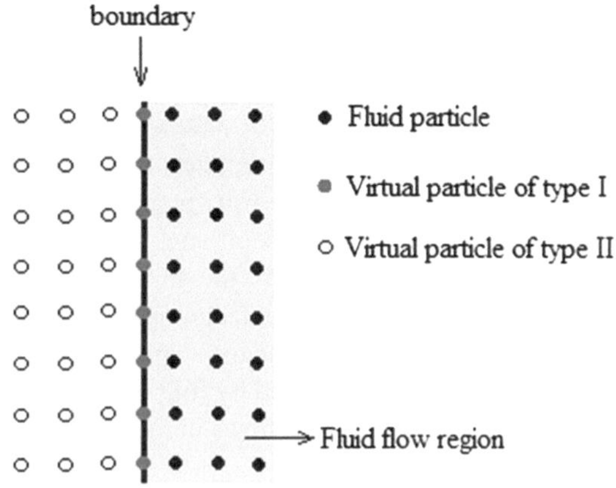

Fig. 3.10 Schematic illustration of the solid contour region. Arrangement of the virtual particles of type I (a line on the contour) and of the virtual particles of type II (in an extended area beyond the domain)

the centres of both particles ($\mathbf{F}_{i\text{v}}$):

$$\mathbf{F}_{i\text{v}} = \begin{cases} \Psi\left[\left(\dfrac{r_0}{|r_{i\text{v}}|}\right)^{n_1} - \left(\dfrac{r_0}{|r_{i\text{v}}|}\right)^{n_2}\right]\dfrac{X_{i\text{v}}}{|r_{i\text{v}}|^2}, \text{ if } \dfrac{r_0}{|r_{i\text{v}}|} \leq 1 \\ 0, \text{ if } \dfrac{r_0}{|r_{i\text{v}}|} > 1. \end{cases} \tag{3.113}$$

where
Ψ is a parameter that depends on each problem and must be of the same order of magnitude as the square of the highest flow velocity
r_0 is the cut-off distance (dependent on the studied problem)
i and v are subscripts that refer to the fluid and virtual particle, respectively
$|r_{i\text{v}}|$ is the distance between the centres of mass of a fluid and a virtual particle
n_1 and n_2 are parameters that depend on each problem (they are usually 12 and 4, respectively)
$X_{i\text{v}}$ is the difference between the position vectors of the virtual particle and the fluid particle.

The sum of the force contributions of each virtual particle on the fluid particle will be the resultant force that will keep the latter within the domain.

Regarding the mirror particles, when a fluid particle is close to the boundary, a mirror particle is placed symmetrically outside the wall. If a fluid particle is near the corner, in addition to the two mirror particles placed opposite the wall, a third mirror particle opposite the corner should be placed. The mirror particles have the same density and pressure as the corresponding inner ones but opposite velocities [26,72].

The number of mirror particles varies at the time, which brings difficulties to the computational implementation of this technique. When using dummy particles, they are regularly distributed outside the domain, parallel to the particles fixed on the boundaries. Dummy particles' velocities are null and remain this form through the simulation. Several layers of dummy particles can be disposed surrounding the solid boundaries. The number of dummy particle layers is decided from the radius of the compact support (so that kernel truncation of the particles within the domain does not occur). This technique is used to implement Neumman boundary condition (the pressure on a dummy particle is identical to the corresponding pressure on the boundary particles fixed on the wall) [24,26].

The techniques presented in this subsection are widely applied in SPH. However, fictitious particles are a purely computational solution, and artificial repulsive forces technique contains concepts of molecular dynamics applied to real problems in the continuum scale. Thus, both are not recommended to be used in continuum mechanics. The former has no physical foundation to be used, and the latter presents a conceptual contradiction, mixing concepts of two different scales (molecular and continuum).

3.5.2 Dynamic Boundary Conditions

In this boundary treatment technique, dynamic particles are fixed on the boundaries and exert a force on the fluid particles when approaching them. The dynamic particles obey the equations of conservation (mass, momentum balance and energy) and Tait equations (the last for the prediction of dynamic pressures), as well as the fluid particles [63].

The dynamic particles remain fixed in their initial positions (fixed boundaries) or can move according to some externally imposed function (when there are gates or wavemakers, for example) [52].

Figure 3.11 shows the interaction between the dynamic particles fixed on the contour (empty circles arranged in a staggered manner) and a fluid particle that approaches the boundary,

When a fluid particle approaches the solid boundary, the density and pressure of the dynamic particles positioned at the boundaries increase. This results in an increase in the magnitude of the repulsive force exerted on the fluid particle [73]. As a consequence, the fluid particles are maintained within the domain.

This boundary treatment technique, besides employing fictitious particles, presents an undesirable and unphysical behaviour in the repulsion mechanism. Particularly, when a thin layer of water is travelling parallel to the bottom, the hydrostatic pressure of the water column is low in comparison to the repulsive force exerted by the dynamic particles. A gap can be observed between this thin layer of fluid and the bottom [52]. Proposals have been made to solve this problem. One of them is not to update the pressure values on the dynamic particles at each step of the simulation time.

3.5.3 Reflective Boundary Conditions

A model for collision treatment in particle methods aims to define how the particles interact with the contours, which prevent direct contact with the environment. House and Keyser [74] present the description of the collision model.

In this boundary treatment technique, a collision detection and response algorithm, such as the one presented by Fraga Filho [75], bring the particles back into the domain. In a real situation, fluid particles cannot escape due to the presence of the boundaries (non-penetration condition).

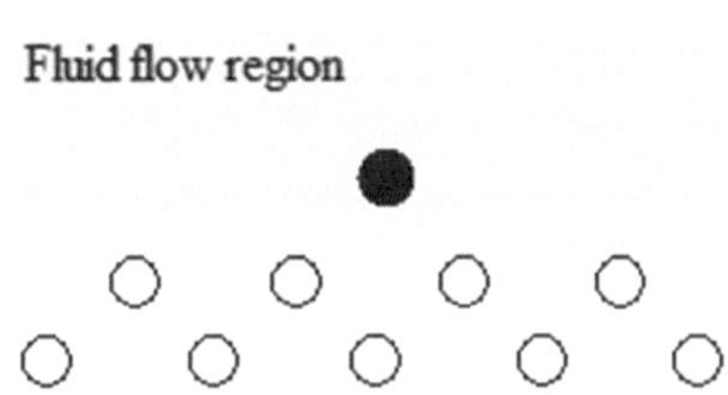

Fig. 3.11 Interaction between the dynamic particles fixed on the contour and a fluid particle that approaches the border (full black circle)

Immediately after collision detection, the particle is reflected; if the particle has not yet returned into the domain, the collision detection and response procedures are repeated until the particle returns to the flow region.

The first step towards obtaining the input data for the algorithm is to solve the momentum conservation equation, which provides the acceleration of the particles at each time step. Following this, the final positions of the centre of mass of the particles and their velocities are obtained using an integration method. The simplest integration method that can be employed is the first-order Runge–Kutta technique (Euler's method). At each time step, the position of the particle is updated using its current derivative (velocity) multiplied by the time interval, and the velocity for the next iteration is given by its current derivative (acceleration calculated using the momentum equation) multiplied by the time step. Both the acceleration and the velocity of the particle are considered to be constant during each time step, and the particle moves linearly along the direction of the velocity (when using Euler's integration method). The positions of the centres of mass of the particles at the initial and final instants of time of the iteration (t_o and ($t_o + \Delta t$), respectively) and the velocities of the particles at each time step form the input data for the algorithm.

The collision detection and response algorithm, based on Newton's restitution law and foundations of analytic geometry, brought the particles back into the domain. The step-by-step algorithm is presented in the following page. The full description and validation can be found in [75].

In the collisions of the fluid particles against the solid walls (considered as being well-defined planes), Newton's restitution law is respected and changes in velocity depend on two different parameters—the coefficient of friction (CF), and the energetic coefficient of restitution due to the elasticity (CR).

Algorithm. Physical reflective boundary conditions: collision detection and response

Input*: Geometry (geometric plane equations), initial and final positions of the centres of mass, and particle velocities

1. **for** each particle in the domain **do**
2. Calculate the distances between the centres of mass of the particles at the end of each iteration (the end of the trajectories) and each plane
3. Check for possible collision planes
4. Identify the plane of collision using the criterion of the smallest trajectory of the particle
5. Reflect the centre of mass of the particle based on the collision plane identified
6. Perform a correction of particle velocity
7. Verify that the particle has returned to the interior of the domain after reflection
8. **if** particle has not returned to the domain **then**
9. Repeat steps 2 to 7 until particle returns to the domain

Output: positions of the centres of mass and particle velocities at the end of each iteration

*At each iteration.

The coefficient of friction is a number from zero to 1.0 giving the fraction of the magnitude of the tangential velocity lost during the collision. A higher value of CF leads to a greater loss of tangential velocity, corresponding to a rougher surface. When CF has the value 1.0, the tangential velocity of the particle is completely wiped out. Friction also produces the rotational movement of the particle.

The elasticity is related to the amount of energy returned to the particle after the collision. The fraction of the returned kinetic energy is given by the coefficient of restitution, CR. Values of CR near 0 show highly inelastic collisions, whereas CR equal to 1.0 indicates a completely elastic collision. Elasticity will produce an effect on the velocity of the particle in the direction normal to the contact point.

Immediately after the particle collision against the plane by the algorithm, the components of the velocities of the particle provided the SPH method are corrected. The magnitude of the velocity component perpendicular to the collision plane immediately after the impact is given by Eq. (3.114):

$$\boxed{(\mathrm{v}_{\mathrm{col}})_N = \mathrm{CR}\,(\mathrm{v}_{\mathrm{p}})_N} \tag{3.114}$$

The friction effect can simply be added to the model implemented in [75], as explained below.

In the tangential direction, we have:

$$\boxed{(\mathrm{v}_{\mathrm{col}})_T = (1.0 - \mathrm{CF})\;(\mathrm{v}_{\mathrm{p}})_T} \tag{3.115}$$

where

$(\mathrm{v}_{\mathrm{col}})_N$ and $(\mathrm{v}_{\mathrm{p}})_N$ are the magnitude of the component of the velocity of the particle perpendicular to the collision plane before and immediately after the collision, respectively

$(\mathrm{v}_{\mathrm{col}})_T$ and $(\mathrm{v}_{\mathrm{p}})_T$ are the magnitude of the tangential component of the velocity before and immediately after the collision, respectively

CR is the coefficient of restitution of kinetic energy

CF is the coefficient of friction.

After the collision, the magnitude and sense of the velocity of the particle and the coordinate of the centre of mass perpendicular to the boundary are altered. The sense of the velocity component parallel to the boundary remained unaltered.

Figure 3.12 shows the behaviour of the velocities of particles after the collision against a plane.

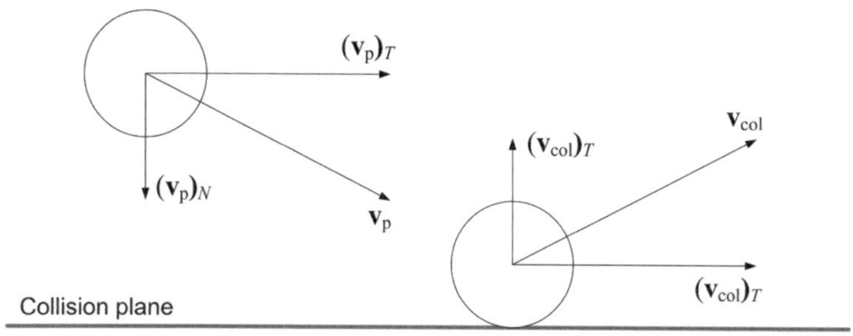

Fig. 3.12 The velocities of particles before the impact against the plane (A) and immediately after it (B)

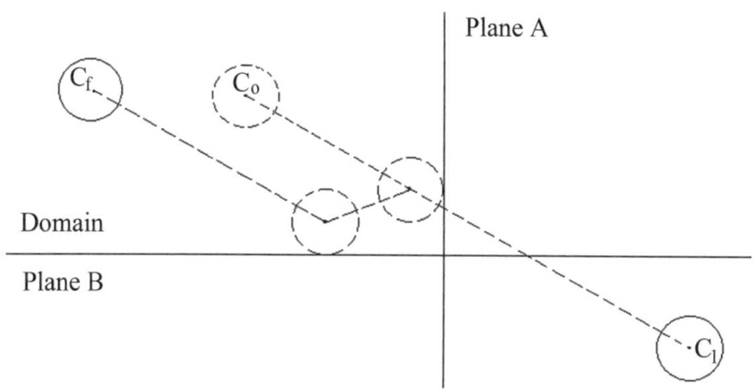

Fig. 3.13 Collisions experienced by a particle in a time step

The coordinate of the centre of mass of the particle perpendicular to the plane must be corrected, as shown in Eq. (3.116):

$$\boxed{(C_f)_N = (C_1)_N + (1.00 + CR)(r_p - d)} \tag{3.116}$$

where

$(C_1)_N$ and $(C_f)_N$ are the ordinates of the mass centre position at the time instant $(to + \Delta t)$, in a movement without obstacles, before and after the reflection, respectively
r_p is the particle radius
d is the distance between the centre of mass of the particle and the collision plane.

Figure 3.13 shows the initial and final positions, C_o and C_f, of the centre of mass of a particle after successively colliding with two planes (A and B) in a numerical iteration. The point C_1 is the final position that would be achieved by the particle mass centre if there were no walls delimiting the domain.

This technique is in accordance with classical mechanics and continuum mechanics. It is not just a computational technique implemented on the continuum from analogies with laws of molecular dynamics.

In complex geometries, the simplest solution is to divide the whole boundary into multiple lines (in 2-D) or small sequential planes (in 3-D). The contour is defined by mathematical equations that represent the lines (or planes); and finding the intersection of a particular line, or plane, with another is a necessary task. If there is a curve, it is possible to approximate it by a series of lines.[1] The more complex the geometry, the more robust the collision detection and response algorithm will be[2].

At the end of this subsection, it is necessary to comment on the dependence between achieving good results in the treatment of collisions, performed by the collisions detection and response algorithm, and the input data. The convergence and accuracy of the numerical results (accelerations of the particles) provided by the SPH solution of the momentum conservation equation and the temporal integration method used to perform the evolution of the system of particles in time are directly related to the results of the reflections performed by the algorithm presented in [75]. In other words, the accuracy in the final positions of the particles, after their reflections and return into the domain, depends on the quality of the input data (provided by the temporal integration performed in the SPH method).

3.5.4 Open Periodic Boundary Conditions

In some problems, as channels fluid flows, the periodic open boundary condition are commonly used. In a simple way, in this boundary treatment particles near an open lateral boundary interact with the particles near the complimentary open lateral boundary on the other side of the domain, Goméz-Gesteira et al. [52]. Figure 3.14 shows, schematically, the domain of influence of a particle '*i*' (extended to the region near to other lateral boundary).

In the periodic open boundary conditions, the particles near the other lateral boundary (within the extended area of influence of the particle '*i*') influence the physical properties the Lagrangian element of reference. In this technique, the particles exiting the computational domain are simply reinserted at the entry, carrying the same properties (velocity, density, pressure and others), Hosseini and Feng [45]. The positions of the particles in space must be updated considering the geometry simulated (the particles moving beyond the lateral boundary should be conducted to the other side of the simulation area as if the two sides were connected).

[1]Barker et al. [76] presented a new technique (the reflected ghost-particle boundaries) in which used B-splines and curved geometries to boundaries; however, utilising fictitious particles in the reflections of the fluid particles in the domain

[2]Reflective techniques also has been used in simulations of collisions against obstacles, as showed by Wu et al. [77] in a fast particle-boundary collision method processed on GPU

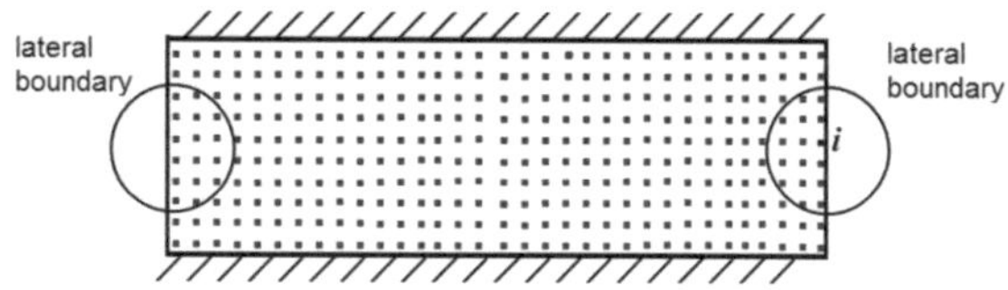

Fig. 3.14 Periodic open boundary conditions. The domain of influence of the particle 'i' is extended to the complimentary open lateral boundary

Tafuni et al. [78] presented computational open boundary conditions, introduced in the open-source SPH code DualSPHysics,[3] using buffer layers and ghost particles.

Leroy [79] presented a boundary technique that discretizes the boundary through vertex particles and segments. The masses of the vertex particles belonging to open boundaries are allowed to evolve over time as a function of the desired input/output mass flow through the open boundary segments. The vertex particles are used to create/delete fluid particles, which is done by setting a minimum and a maximum value for their masses, proportional to the mass of the free particles. The open boundaries algorithm was tested on a laminar plane Poiseuille flow.

3.5.5 Additional Remarks

Besides the boundary treatment techniques here presented, there are others described in the literature. The use of fictitious, virtual, ghost or dynamic particles, as well as the artificial repulsive forces (analogous to those that exist in the molecular scale), should be avoided. These techniques, widely employed in problem-solving using Lagrangian particle methods, are inadequate to be implemented in the continuum scale. Some of them are purely computational boundary treatment techniques, and others mix concepts of molecular and continuum scales, violating the continuum theory. Physically based techniques (which respect the laws of the continuum mechanics) should be employed.

As said in the beginning of this subject, an appropriate boundary treatment technique is one of the greatest challenges found in the Smoothed Particle Hydrodynamics method. Even knowing the difficulties related to obtaining consistent numerical solutions using the SPH method (and that have often been achieved using contour treatment techniques that do not have a physical foundation in the continuum domain), the author is sure that, in the course of time and with the progress of researches, these techniques will be replaced by others more appropriate to the continuum scale in which the methods of particles have been defined, as well as to the physical laws that must be respected.

[3] Available for download at http://dual.sphysics.org/

References

1. Lucy, L.B.: Numerical approach to testing the fission hypothesis. Astron. J. **82**, 1013–1024 (1977)
2. Gingold, R.A., Monaghan, J.J.: Smoothed particle hydrodynamics: theory and application to non-spherical stars. Mon. Not. R. Astron. Soc. **181**, 375–389 (1977)
3. Liu, G.R., Liu, M.B.: Smoothed Particle Hydrodynamics: a Meshfree Particle Method. World Scientific, Singapore (2003)
4. Liu, M.B., Liu, G.R.: Smoothed particle hydrodynamics (SPH): an overview and recent developments. Arch. Comput. Methods Eng. **17**, 25–76 (2010)
5. Monaghan, J.J.: Smoothed particle hydrodynamics. Annu. Rev. Astron. Astrophys. **30**, 543–574 (1992)
6. Quinlan, N.J., Basa, M., Lastiwka, M.: Truncation error in mesh-free particle methods. Int. J. Numer. Methods Eng. **66**, 2064–2085 (2006)
7. Vaughan, G.L., Healy, T.R., Bryan, K.R., Sneyd, A.D., Gorman, R.M.: Completeness, conservation and error in SPH for fluids. Int. J. Numer. Methods Fluids **56**, 37–62 (2008)
8. Morris, J.P., Fox, P.J., Zhu, Y.: Modeling low reynolds number incompressible flows using SPH. J. Comput. Phys. **136**, 214–226 (1997)
9. Liu, M.B., Liu, G.R., Lamb, K.Y.: Constructing smoothing functions in smoothed particle hydrodynamics with applications. J. Comput. Appl. Math. **155**, 263–284 (2003)
10. Hongbin, J., Xin, D.: On criterions for smoothed particle hydrodynamics kernels in stable field. J. Comput. Phys. **202**, 699–709 (2005)
11. Dehnen, W., Hossam, A.: Improving convergence in smoothed hydrodynamics simulations without pairing instability. Mon. Not. R. Astron. Soc. **425**, 1068–1082 (2012)
12. Yang, X.F., Peng, S.L., Liu, M.B.: A new kernel function for SPH with applications to free surface flows. Appl. Math. Modell. **38**, 3822–3833 (2014)
13. Domínguez, J.M., Crespo, A.J.C., Gómez-Gesteira, M., Marongiu, J.C.: Neighbour lists in smoothed particle hydrodynamics. Int. J. Numer. Meth. Fluids **67**, 2026–2042 (2011)
14. Viccione, G., Bovolin, V., Pugliese, Carratelli E.: Defining and optimizing algorithms for neighbouring particle identification in SPH fluid simulations. Int. J. Num. Methods Fluids **58**, 625–638 (2008)
15. Rajasekaran, S., Reif, J.: Handbook of parallel computing: models, algorithms and applications. Chapmann and Hall/CRC, Boca Raton, FL (2007)
16. Jahanbakhsh, E., Pacot, O., Avellan, F.: Implementation of a parallel SPH-FPM solver for fluid flows. Numerical Simulation Sci. Technol. **1**, 16–20 (2012). http://zetta.epfl.ch/2012/R-ZETTA-WEB.pdf. Accessed 30 August 2018
17. Gingold, R.A., Monaghan, J.J.: Shock simulation by the particle method SPH. J. Comput. Phys. **52**, 374–389 (1983)
18. Brackbill, J.U., Kothe, D.B., Zemach, C.: A continuum method for modeling surface tension. J. f Comput. Phys. **100**, 335–354 (1992)
19. Müller, M., Charypar, D., Gross, M.: Particle-based fluid simulation for interactive applications. In: Proceedings of the Eurographics/SIGGRAPH Symposium on Computer Animation, San Diego, CA, USA, pp. 154–159 (2003)
20. Fossum, F.: Surface tension in smoothed particle hydrodynamics on the GPU. In: Complex Computer Systems, Specialization Project, NTNU, Norway, 2010. http://www.idi.ntnu.no/elster/master-studs/fossum/fossum-fall2010-proj.pdf. Accessed 28 Jan. 2018
21. Morris, J.P.: Simulating surface tension with smoothed particle hydrodynamics. Int. J. Numer. Methods Fluids **33**, 333–353 (2000)
22. Zhang, M., Zhang, S., Zhang, H., Zheng, L.: Simulating of surface-tension-driven interfacial flow with smoothed particle hydrodynamics methods. Comput. Fluids **59**, 61–71 (2012)
23. Ataie-Ashtiani, B., Shobeyri, G., Farhadi, L.: Modified incompressible SPH method for simulating free surface problems. Fluid Dyn. Res. **40**, 637–661 (2008)

24. Lee, B., Park, J., Kim, M., Hwang, S.: Step-by-step improvement of MPS method in simulating violent free-surface motions and impact-loads. Comput. Methods Appl. Mech. Engrg **200**, 1113–1125 (2011)
25. Ramli, M.Z., Temarel, P., Tan M.: Smoothed Particle Hydrodynamics (SPH) method for modelling 2-Dimensional free surface hydrodynamics. In: Guedes Soares C., Shenoi R.A. (eds.) Analysis and Design of Marine Structures V. CRC Press. **8**, 45–52
26. Lee, E.-S., Moulinec, C., Xu, R., Violeau, D., Laurence, D., Stansby, P.: Comparisons of weakly compressible and truly incompressible algorithms for the SPH mesh free particle method. J. Comput. Phys. **227**(18), 8417–8436 (2008)
27. Gabreil, E., Tait, S.J., Shao, S., Nichols, A.: SPHysics simulation of laboratory shallow free surface turbulent flows over a rough bed. J. Hydraul. Res. (2018). https://doi.org/10.1080/00221686.2017.1410732
28. Farhadi, A., Ershadi, H., Emadad, H., Rad, E.G.: Comparative study on the accuracy of solitary wave generations in an ISPH-based numerical wave flume. Appl. Ocean Res. **54**, 115–136 (2016)
29. Liu, M.B., Liu, G.R., Lam, K.Y., Zong, Z.: Smoothed particle hydrodynamics for numerical simulation of underwater explosion. Comput. Mech. **30**(2), 106–118 (2003)
30. Zhu, Z.X., Zou, L., Chen, Z., Wang, A.M., Liu, M.B.: An improved SPH model for multiphase flows with large density ratios. Int. J. Numer. Methods Fluids **86**(2), 167–184 (2018)
31. Jian, W., Liang, D., Shao, S., Chen, R., Liu, X.: SPH Study of the evolution of water-water interfaces in dam break flows. Nat. Hazards **78**(1), 531–553 (2015)
32. Fraga Filho, C.A.D., Chacaltana, J.T.A., Pinto, W.J.N.: Meshless Lagrangian SPH method applied to isothermal lid-driven cavity flow at low-re numbers. Comp. Part. Mech. (2018). https://doi.org/10.1007/s40571-018-0183-x
33. Monaghan, J.J.: SPH compressible turbulence. Mon. Not. R. Astron. Soc. **335**(3), 843–852 (2002). https://doi.org/10.1046/j.1365-8711.2002.05678.x
34. Chen, S., Holm, D.D., Margolin, L.G., Zhang, R.: Direct numerical simulations of the Navier-Stokes Alpha model. Phys. D Nonlinear Phenom. **133**(1–4) 66, 83 (1999). https://doi.org/10.1016/S0167-2789(99)00099-8
35. Mansour, J.: SPH and α-SPH: applications and analysis. Ph.D. Thesis, Monash University, Australia (2007)
36. Dalrymple, R.A., Rogers, B.D.: Numerical modeling of water waves with the SPH method. Coast. Eng. **53**, 141–147 (2006)
37. Violeau, D., Issa, R.: Numerical modelling of complex turbulent free-surface flows with the SPH method: an overview. Int. J. Numer. Methods Fluids **53**(2), 277–304 (2007)
38. Monaghan, J.J.: A turbulence model for smoothed particle hydrodynamics. Eur. J. Mech. B/Fluids **30**, 360–370 (2011)
39. Robinson, M., Monaghan, J.J.: Direct numerical simulation of decaying two-dimensional turbulence in a no-slip square box using smoothed particle hydrodynamics. Inte. J. Numer. Methods Fluids **70**, 37–55 (2012)
40. Adami, S., Hu, X.Y., Adams, N.A.: Simulating three-dimensional turbulence with SPH. In: Center for Turbulent Research, Proceedings of the Summer Program, Stanford University (2012)
41. Mayrhofer, A., Laurence, D., Rogers, B.D., Violeau, D.: DNS and LES of 3-D wall-bounded turbulence using smoothed particle hydrodynamics. Comput. Fluids **115**, 86–97 (2015)
42. De Padova, D., Dalrymple, R.A., Mossa, M.: Analysis of the artificial viscosity in the smoothed particle hydrodynamics modelling of regular waves. J. Hydrau. Res. (2014). https://doi.org/10.1080/00221686.2014.932853
43. Monaghan, J.J.: On the problem of penetration in particle methods. J. Comput. Phys. **82**, 1–15 (1989)
44. Paiva, P., Petronetto, F., Lewiner, T., Tavares, G.: Particle-based viscoplastic fluid/solid simulation. Compu. Aided Des. **41**, 306–314 (2009)

45. Hosseini, S.M., Feng, J.J.: Pressure boundary conditions for computing incompressible flows with SPH. J. Comput. Phys. **230**, 7473–7487 (2011)
46. Monaghan, J.J.: SPH without tensile instability. J. Comput. Phys. **159**, 290–311 (2000)
47. Lobovský, L., Kren, J.: Smoothed particle hydrodynamics modelling of fluids and solids. Applied and Computational Mechanics **1**, 521–530 (2007)
48. Swegle, J. Conservation of momentum and tensile instability in particle methods. Sandia Report, SAND2000-1223 (2000). https://inis.iaea.org/collection/NCLCollectionStore/Public/32/026/32026572.pdf. Accessed 30 July 2018
49. Khayyer, A., Gotoh, H.: Enhancement of stability and accuracy of the moving particle semi-implicit method. J. Comput. Phys. **230**, 3093–3118 (2011)
50. Chen, J.K., Beraun, J.E., Jih, C.J.: An improvement for tensile instability in smoothed particle hydrodynamics. Comput. Mech. **23**, 279–287 (1999)
51. Yang, X., Liu, M., Peng, S.: Smoothed particle hydrodynamics modeling of viscous liquid drop without tensile instability. Computers & Fluids **92**, 199–208 (2014)
52. Gómez-Gesteira, M., Rogers, B.D., Crespo, A.J.C., Dalrymple, R.A., Narayanaswamy, M., Dominguez, J.M.: SPHysics–development of a free surface fluid solver–part 1: theory and formulations. Comput. Geosci. **48**, 289–299 (2012)
53. Courant, R., Friedrichs, K., Lewy, H.: On the partial difference equations of mathematical physics. IBM J. **11**, 215–234 (1967)
54. Monaghan, J.J., Koss, A.: Solitary waves on a Cretan beach. J. Waterw. Port Coast. Ocean Eng. **125**, 145–155 (1999)
55. Chen, J.K., Beraun, J.E., Carney, T.C.: A corrective smoothed patricle method for boundary value problems in heat conduction. Int. J. Numer. Methods Eng. **46**, 231–252 (1999)
56. Dilts, G.A.: Moving-least-squares-particle hydrodynamics-I. Consistency and stability. Int. J. Numer. Meth. Eng. **44**, 1115-1155 (1999)
57. Colagrossi, A., Landrini, M.: Numerical simulation of interfacial flows by smoothed particle hydrodynamics. J. Comput. Phys. **191**, 448–475 (2003)
58. Monaghan, J.J.: Simulating free surface flows with SPH. J. Comput. Phys. **110**, 399–406 (1994)
59. Lobovský, L., Groenenboomb, P.H.L.: Smoothed particle hydrodynamics modelling in continuum mechanics: fluid-structure interaction. Appl. Comput. Mech. **3**, 101–110 (2009)
60. Yildiz, M., Rook, R.A., Suleman, A.: SPH with the boundary tangent method. Int. J. Numer. Methods Eng. **77**, 1416–1438 (2009)
61. Fourtakas, G., Vacondio, R., Rogers, B.D.: On the approximate zeroth and first-order consistency in the presence of 2-D irregular boundaries in SPH obtained by the virtual boundary particle methods. Int. J. Numer. Methods Fluids **78**(8), 475–501 (2015)
62. Adami, S., Hu, X.Y., Adams, N.A: A generalized wall boundary condition for smoothed particle hydrodynamics. J. Comput. Phys. **231**, 7057–7075 (2012)
63. Crespo, A.J.C., Gómez-Gesteira, M., Dalrymple, R.A.: Boundary conditions generated by dynamic particles in SPH methods. CMC Comput. Mat. Cont. **5**(3), 173–184 (2007)
64. Kulasegaram, S., Bonet, J., Lewis, R.W., Profit, M.: A variational formulation based contact algorithm for rigid boundaries in two-dimensional SPH applications. Comput. Mech. **33**, 316–325 (2004)
65. Ferrand, M., Laurence, D.R., Rogers, B.D., Violeau, D., Kassiotis, C.: Unified semi-analytical wall boundary conditions for inviscid, laminar or turbulent flows in the meshless SPH method. Int. J. Numer. Methods Fluids **71**(4), 446–472 (2013)
66. Leroy, A., Violeau, D., Ferrand, M., Kassiotis, C.: Unified semi-analytical wall boundary conditions applied to 2-D incompressible SPH. J. Comput. Phys. **261**, 106–129 (2014)
67. Mayrhofer, A., Ferrand, M., Kassiotis, C., Violeau, D., Morel, F.: Unified semi-analytical wall boundary conditions in SPH: analytical extension to 3-D. Numer. Algorithms **68**, 15–34 (2015)

68. Macià, F., Gonzales, L.M., Cercos-Pita, J.L., Souto-Iglesias A.A: Boundary Integral SPH Formulation - Consistency and Applications to ISPH and WCSPH. Progr. Theor. Exp. Phys. **128**(3), 439–462 (2012)
69. Chiron, L., de Leffe, M., Oger, G., Le Touzé, D.: Fast and accurate SPH modelling of 3D complex wall boundaries in viscous and non viscous flows. Comp. Phy. Commun. https://doi.org/10.1016/j.cpc.2018.08.001
70. Valizadeh, A., Shafieefar, M., Monaghan, J.J., Neyshabouri, S.A.A.S.: Modeling two-phase flows using SPH method. J. Appl. Sci. **8**(21), 3816–3826 (2008)
71. Rapaport, D.C.: The Art of Molecular Dynamics Simulation, 3rd edn. Cambridge University Press, UK (2004)
72. Dong, T., Shunliang, J.: Comparisions of mirror and static boundary conditions in incompressible smoothed particle hydrodynamics. In: Proceedings of International Conference on Computational and Information Sciences, Chengdu, (2010). https://doi.org/10.1109/ICCIS.2010.299
73. Fraga Filho, C.A.D., Chacaltana, J.T.A.: Boundary treatment techniques in smoothed particle hydrodynamics: implementations in fluid and thermal sciences and results analysis. Interdisciplinary Journal of Engineering Research—RIPE. In: Proceedings of the XXXVII Iberian Latin American Congress on Computational Methods in Engineering—CILAMCE 2016, Brasília-DF, 2016. http://periodicos.unb.br/index.php/ripe/article/view/23421. Accessed 10 April 2017
74. House, D.H., Keyser, J.C.: Foundations of Physically Based Modeling & Animation. CRC Press. Taylor & Francis Group, Boca Raton, Florida, USA (2017)
75. Fraga Filho, C.A.D.: An algorithmic implementation of physical reflective boundary conditions in particle methods: Collision detection and response. Phys. Fluids **29**, 113602 (2017). https://doi.org/10.1063/1.4997054
76. Barker, D.J., Brito-Parada, P., Neethling, S.J.: Application of B-splines and curved geometries to boundaries in SPH. Int. J. Numer. Methods Fluids **76**, 51–68 (2014)
77. Wu, J., Zhang, F., Shen, X.: GPU-Based fluid simulation with fast collision detection on bBoundaries. International Journal of Modeling, Simulation, and Scientific Computing **3**(1), 1240003 (2012). https://doi.org/10.1142/S179396231240003X
78. Tafuni, A., Dominguez, J. M., Vacondio, R., Sahin, I., Crespo, A.J.C.: Open boundary conditions for large-scale SPH simulations. In: Proceedings of the 11th SPHERIC International Workshop, June 13-16, Munich, Germany, 2016. https://scholar.google.com/citations?user=mEr7crYAAAAJ&hl=en. Accessed 20 July 2018
79. Leroy, A.: Un nouveau modèle SPH incompressible: vers l'application à des cas industriels. Université Paris-Est, France, 2014. http://www.theses.fr/2014PEST1065. Accessed 20 July 2018

Applications in Continuum Fluid Mechanics and Transport Phenomena

4

Abstract

This chapter brings some applications of the SPHd method in transport phenomena and continuum fluid mechanics problems. The physical-mathematical modellings, the boundary treatment techniques implemented and the results achieved are presented and discussed. Four cases are studied: heat diffusion in a flat plate, still fluid at rest within an immobile reservoir, dam breaking and oil spreading on a calm sea. In the studies, new solutions to problems with results already existing (still fluid within an immobile reservoir and dam breaking) or not existing in the literature (oil spreading on a calm sea) are proposed taking care to avoid the use of molecular concepts in the macroscopic scale of the continuum. At the end, concluding remarks about each case studied are listed.

Keywords

Continuum fluid mechanics · Transport phenomena
Heat diffusion in a flat plate · Fluid at rest · Dam breaking · Oil spreading

4.1 Procedure Employed in Problem-Solving

In each problem solution presented in this chapter, the procedure shown in the following flowchart has been performed (Fig. 4.1).

In the context of the scientific method, presented in Sect. 1.2, the improvement of models previously presented in the literature has been sought, especially with regard to the implementation of boundary treatment techniques. The next sections bring the problems modelling, numerical simulations, results analysis, discussions and final comments.

C. A. Dutra Fraga Filho, *Smoothed Particle Hydrodynamics*,
https://doi.org/10.1007/978-3-030-00773-7_4

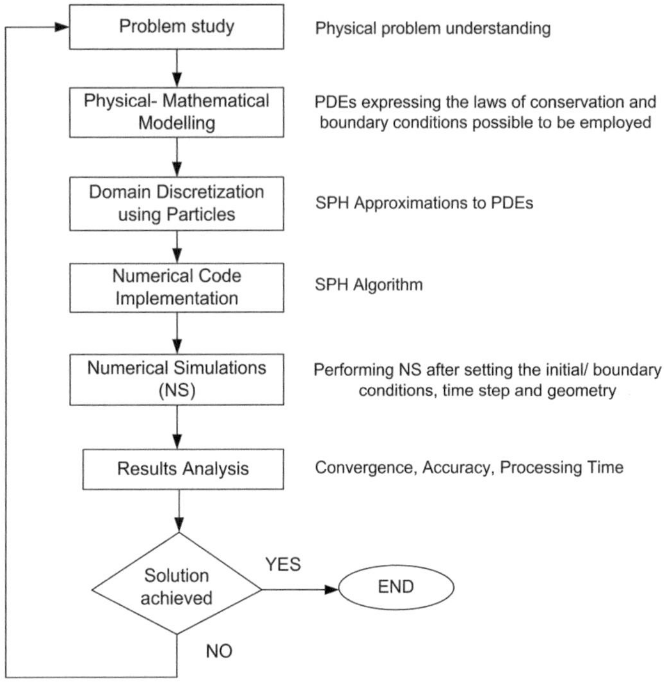

Fig. 4.1 Procedure employed to obtaining a numerical solution to a physical/engineering problem

4.2 Heat Diffusion in a Homogeneous Flat Plate

Analysing the heat diffusion from the molecular microscopic point of view, free electrons packed closely together move (and vibrate) throughout the matter. They pick up kinetic energy in collisions with hot cores and loose it again in collision with cooler cores. In this process, thermal energy moves through the matter. Molecular dynamics models for heat diffusion can be found in Maruyama [1] and Wang and Xu [2].

In the Lagrangian SPH modelling employed in this work, the heat diffusion occurs due to differences of temperature between the particles at rest in the domain. The electrons vibration is not seen because the study is focused on the heat transfer among the macroscopic particles at the continuum domain.

4.2.1 Physical-Mathematical Modelling

The energy conservation equation, Eq. (2.3), in the absence of pressure and energy dissipation fields, for an incompressible material, is written as follows:

$$\boxed{\frac{1}{\alpha}\frac{DT}{Dt} = \left(\frac{\partial^2 T}{\partial x^2} + \frac{\partial^2 T}{\partial y^2}\right) + \frac{\dot{q}}{K}} \tag{4.117}$$

$$\alpha = \frac{K}{\rho c_v} \tag{4.118}$$

where

α is the thermal diffusivity
K is the thermal conductivity
c_v is the specific heat at constant volume.

When the material is homogeneous, with uniform properties, and there are no heat sources, Eq. (4.117) is simplified:

$$\boxed{\frac{1}{\alpha}\frac{DT}{Dt} = \left(\frac{\partial^2 T}{\partial x^2} + \frac{\partial^2 T}{\partial y^2}\right)} \tag{4.119}$$

At the steady state, Eq. (4.119) becomes:

$$\boxed{\left(\frac{\partial^2 T}{\partial x^2} + \frac{\partial^2 T}{\partial y^2}\right) = 0} \tag{4.120}$$

The whole domain was at a temperature T_0 in the initial instant; that is, the initial temperature was uniform. In the boundaries, null temperatures were prescribed (Dirichlet boundary conditions), except in the lower one, which was assigned a temperature of 100 °C.

The initial and boundary conditions are presented below.

The initial condition is:

$T(x, y) = T_0 = 0\,°C$, $(0 < x < 1$ and $0 < y < 1)$

The boundary conditions are:

$T(0, y) = 0\,°C$, $(0 \le y \le 1)$,
$T(1, y) = 0\,°C$, $(0 \le y \le 1)$,
$T(x, 0) = T_S = 100\,°C$, $(0 \le x \le 1)$,
$T(x, 1) = 0\,°C$, $(0 \le x \le 1)$.

Figure 4.2 shows the flat plate, the initial and boundary conditions.

The SPH method was used for the solution of the heat diffusion equation, from the transient regime, until the steady state was reached.

The calculation of the Laplacian of temperature for each iteration was done explicitly by solving Eq. (4.121). In Appendix B, there is a detailed deduction of the Laplacian SPH operator.

Fig. 4.2 Homogeneous flat plate with initial and boundary conditions

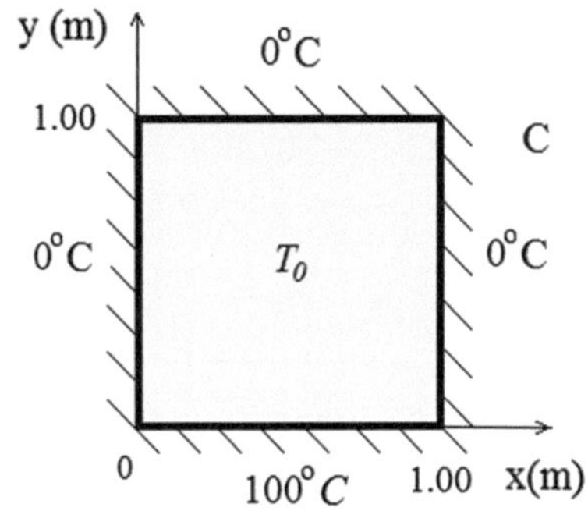

$$\left(\frac{\partial^2 T_i}{\partial x^2} + \frac{\partial^2 T_i}{\partial y^2}\right)^m = 2\alpha_i \sum_{j=1}^{n} \frac{m_j}{\rho_j} \left[T(X_i)^m - T(X_j)^m\right] \frac{\partial W(\mathbf{r}_i - \mathbf{r}_j, h)}{\partial r} \frac{1}{|\mathbf{r}_{ij}|} \tag{4.121}$$

where

T_i is the temperature of the fixed particle
T_j is the temperature of the neighbouring particle
α_i is the thermal diffusivity of the fixed particle
m is the current time step
$|\mathbf{r}_{ij}| = |(X_i - X_j)|$ is the distance between the fixed and a neighbouring particle.

The forecast of temperatures of the particles was performed by Euler's numerical integration method.

$$T^{m+1} = T^m + \left(\frac{\partial^2 T}{\partial x^2} + \frac{\partial^2 T}{\partial y^2}\right)^m \Delta t \tag{4.122}$$

The SPH results have been compared to the solution obtained by series for the temperature distribution in an isotropic and homogeneous flat plate, in the stationary state [3]:

$$T(x, y) = \sum_{N=1}^{\infty} A_N \sin(N\pi x) \sinh[N\pi(y-1)] \tag{4.123}$$

$$A_N = \frac{2T_S}{N\pi} \frac{[(-1)^N - 1]}{\sinh(N\pi)} \tag{4.124}$$

where

T_S is the temperature at the lower boundary
N is the number of points employed in the series
A_N are the coefficients of the series.

The first ninety terms of the series were used in the calculations.

4.2.2 Numerical Simulations

The computational domain consisted of a square region with a side length of 1.00 m representing the sides of the flat plate. The domain was discretized by particles in the form of circles that were tangent in the initial moment and remained stationary throughout the simulation. The mass centres of each particle were laterally separated by a distance $dx = dy = 1/\mathrm{n_p}$, where $\mathrm{n_p}$ is the number of particles arranged on each side of the domain. Five different values of $\mathrm{n_p}$ (50, 60, 70, 80 and 90) have been employed.

In this problem, in particular, artificial repulsive forces were not employed. The virtual particles were defined for the application of the Dirichlet boundary conditions only.

Different initial particle configurations were defined in order to verify the effect of the number of particles used in the discretization of the domain in the results of the simulations. Table 4.1 shows the number of particles per side of the domain and the respective lateral distances between the centres of mass.

Figure 4.3 presents a flowchart of the algorithm employed for simulations. FORTRAN source files for this problem are available at the end of this book.

Table 4.1 Initial configurations of particles and lateral distances of the centres of mass

No. of particles per side	Lateral distance (m)
50	0.0200
60	0.0167
70	0.0143
80	0.0125
90	0.0111

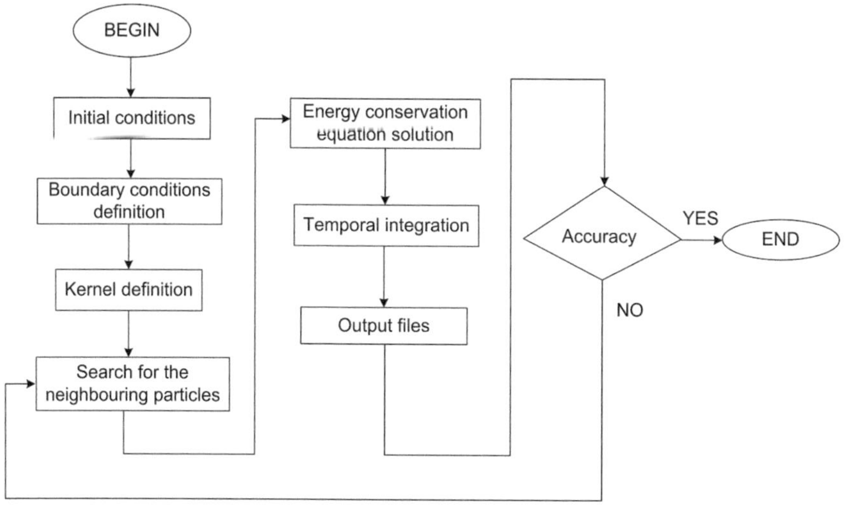

Fig. 4.3 Algorithm implemented for diffusion in a homogeneous flat plate

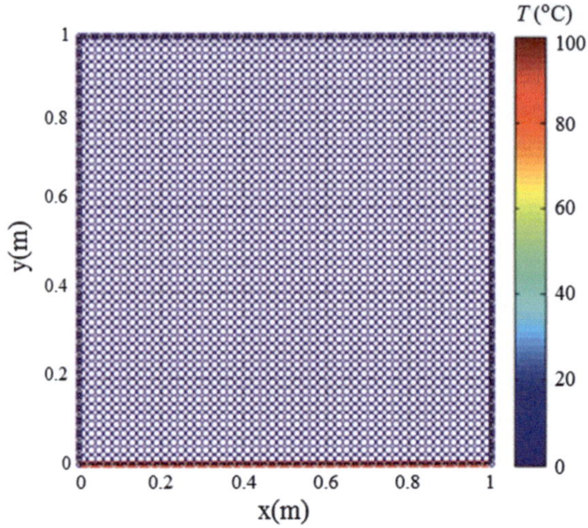

Fig. 4.4 Initial setup of 50×50 particles distributed inside the domain and 100 particles on each side of the plate

In the defining of the boundaries, a line of particles has been fixed at the contours at a ratio of 02 boundary particles for each real particle of the domain, and the temperatures of the sides of the plate (Dirichlet boundary conditions) were attributed to them. The temperatures of the particle temperatures on the edges did not suffer interpolations over time.

The particles inside the domain have had their temperatures initialized to 0 °C, and physical properties have been allocated for them ($\alpha = 1.0\,\text{m}^2/\text{s}$, with $K = 1.0\,\text{J/s}\,^\circ\text{C}\,\text{m}$, $\rho = 1.0\,\text{kg/m}^3$ and $c_v = 1.0\,\text{J/kg}\,^\circ\text{C}$, held constant for all simulations). The influence domain (with radius kh) was defined as $2.50\,dx$, invariable during simulations. The time step was 1.0×10^{-5} s.

Figure 4.4 shows a initial setup of 50 particles per side of the domain and the boundary particles on the edges (Dirichlet boundary conditions implemented).

The spatial points in which the solution by series have been obtained were coincident with the centres of mass of the particles (SPH method). Four different kernels were employed in the SPH simulations: cubic spline, Lucy's quartic, new quartic and quintic spline, presented in Sect. 3.2.6.

The termination of the simulations occurred when the steady state was reached. The stopping criterion was defined by an error less than $1.0 \times 10^{-6}\,^\circ\text{C}$, for the temperature in each particle in the domain, calculated as follows:

$$\left|T^{m+1} - T^m\right| < \text{error} \tag{4.125}$$

After reaching the steady state, the effect of the contours on the solution obtained by SPH was clearly noticed. In the border regions, the kernel consistency is not guaranteed. The kernel truncation at the boundaries led to the greatest differences between particle temperatures, (ΔT), when comparing SPH and series solution at the same spatial points.

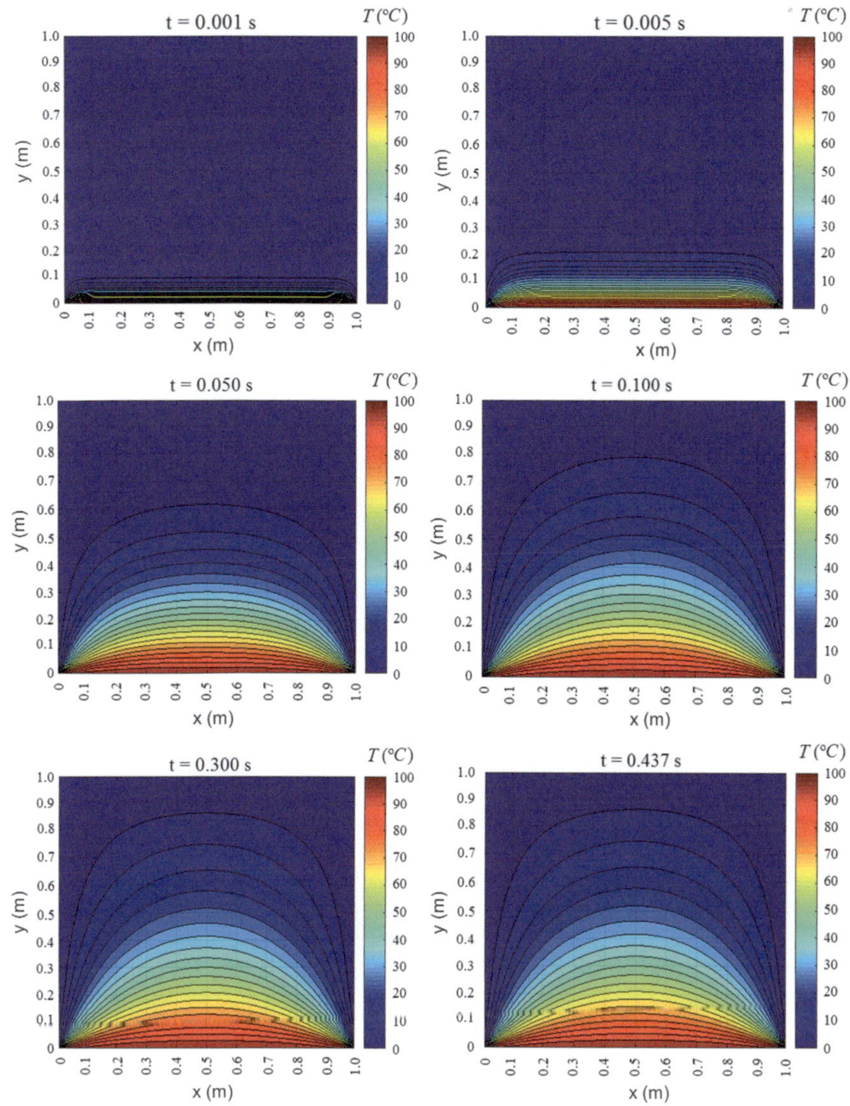

Fig. 4.5 Heat diffusion at different instants of time, until the steady state is reached (spline cubic kernel, 2,500 particles)

The physical time for the occurrence of heat diffusion from a transitional regime to steady state, obtaining the specified accuracy, was contained between 0.43124 and 0.45366 s. Figure 4.5 shows the occurrence of heat diffusion in its transitional regime at different instants of time, until steady state is reached.

Table 4.2 presents the largest differences between the temperatures (highest temperature errors) obtained by series solution and SPH. These differences are also expressed in percentage form, calculated from the expression:

Table 4.2 Largest percentage difference between temperatures provided by series and SPH solution

No. of particles	Kernel											
	Cubic spline			Lucy's quartic			New quartic			Quintic spline		
	T_{SPH} (°C)	ΔT (°C)	ΔT_p (%)	T_{SPH} (°C)	ΔT (°C)	ΔT_p (%)	T_{SPH} (°C)	ΔT (°C)	ΔT_p (%)	T_{SPH} (°C)	ΔT (°C)	ΔT_p (%)
50 × 50	17.9739	7.5735	29.65	17.5502	7.9972	31.30	17.0137	8.5337	33.40	13.3070	7.1435	34.93
60 × 60	17.0814	8.5357	33.32	68.6934	10.0822	12.80	16.9479	8.6692	33.84	16.9108	8.7063	33.99
70 × 70	65.8531	12.6761	16.14	64.7027	13.8265	17.61	66.0121	12.5171	15.94	18.3173	7.3419	28.61
80 × 80	62.7139	16.5570	20.88	62.1542	17.1167	21.59	63.3062	15.9647	20.14	15.1684	10.5182	40.95
90 × 90	60.5852	20.0315	24.85	60.2273	20.3904	25.29	61.1515	19.4662	24.15	65.4389	15.1788	18.83

Table 4.3 Positions of the particles where the greatest differences between temperatures occurred

No. of particles	Kernel							
	Cubic spline		Lucy's quartic		New quartic		Quintic spline	
	x(m)	y(m)	x(m)	y(m)	x(m)	y(m)	x(m)	y(m)
50 × 50	3.00×10^{-2}	7.00×10^{-2}	3.00×10^{-2}	1.00×10^{-2}	3.00×10^{-2}	1.00×10^{-2}	3.00×10^{-2}	1.00×10^{-3}
60 × 60	2.50×10^{-2}	5.83×10^{-2}	2.50×10^{-2}	8.33×10^{-2}	2.50×10^{-2}	5.33×10^{-2}	2.50×10^{-2}	5.83×10^{-2}
70 × 70	2.14×10^{-2}	7.14×10^{-3}	2.14×10^{-2}	7.14×10^{-3}	2.14×10^{-2}	7.14×10^{-3}	2.14×10^{-2}	7.14×10^{-3}
80 × 80	9.81×10^{-1}	6.25×10^{-3}	9.81×10^{-1}	6.25×10^{-3}	9.81×10^{-1}	6.25×10^{-3}	1.88×10^{-2}	4.38×10^{-2}
90 × 90	9.83×10^{-1}	5.56×10^{-3}	9.83×10^{-1}	5.56×10^{-3}	9.83×10^{-1}	5.56×10^{-3}	1.67×10^{-2}	5.56×10^{-3}

$$\Delta T_p = \left(\frac{T_{series} - T_{SPH}}{T_{series}} \right) 100 = \left(\frac{\Delta T}{T_{series}} \right) 100 \tag{4.126}$$

where

ΔT_p is the largest percentage difference between temperatures in a defined spatial point
T_{series} is the temperature obtained by series
T_{SPH} is the temperature provided by SPH solution
ΔT is the temperature error.

Table 4.3 shows the positions of the particles where the greatest differences between the temperatures occurred (presented in Table 4.2). From the analysis of these two tables, it was seen that the greatest temperature differences occurred at the lower corners, where there were the highest temperature gradients.

The amount of particles in the domain also had significant influence. By increasing the number of particles in the domain, the distances between them decreased, which has resulted in a reduction of temperature gradients in the inner regions and an elevation of the temperature gradients on the corners, for all kernels utilized. When more particles were added to the domain, it was observed that the relative errors (ΔT), although greater in magnitude, were restricted to a smaller number of spatial positions near the lower corners.

Figure 4.6 shows the behaviour for different particle numbers, for a unique kernel interpolation (quintic spline).

Another variable which was equally important was the kernel used. Regarding the interpolation functions employed, there was an influence of the kernel degree on the distributions of temperatures obtained. The smaller differences of temperatures in regions close to the boundaries have been observed when the kernel of higher degree was utilized (quintic spline), except for the distribution with 3,600 particles.

Figure 4.7 shows the largest relative errors for different combinations of particle numbers and kernels (all in a few positions on the lower corners of the flat plate).

4.2.3 Comments

The implementation of Dirichlet boundary conditions using lines of boundary particles fixed on the edges of the domain led to satisfactory results, especially in regions remote from the contours. There was a good concordance between the temperatures provided by SPH particle method and series solutions.

Near the boundaries, the kernel truncation and particle inconsistency phenomenon were seen. The interpolation of the gradient of the kernel had a lower precision, leading to larger relative errors and the largest differences of temperatures (SPH/analytical).

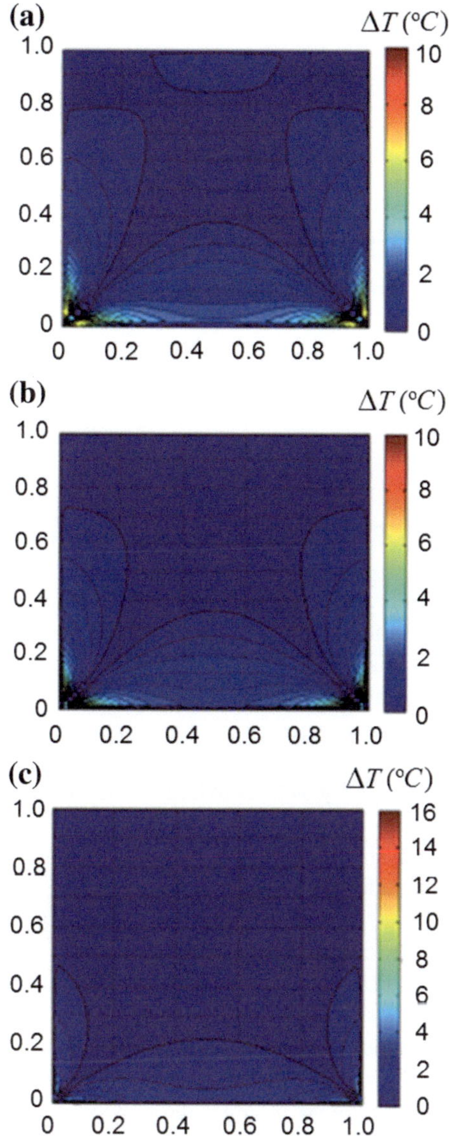

Fig. 4.6 Effect of increasing the particle number in the domain. Solution by using the quintic spline kernel: **a** 2,500, **b** 4,900 and **c** 8,100 particles

The implementation and testing of correction techniques for the temperatures near the boundaries should be carried out in conjunction with an adequate boundary condition treatment, which respects the laws of continuum mechanics. Some methods, such as those presented in the literature (for example, [4–7]), deserve to be implemented and their contributions for the improvement of the results in the regions close to the contours verified.

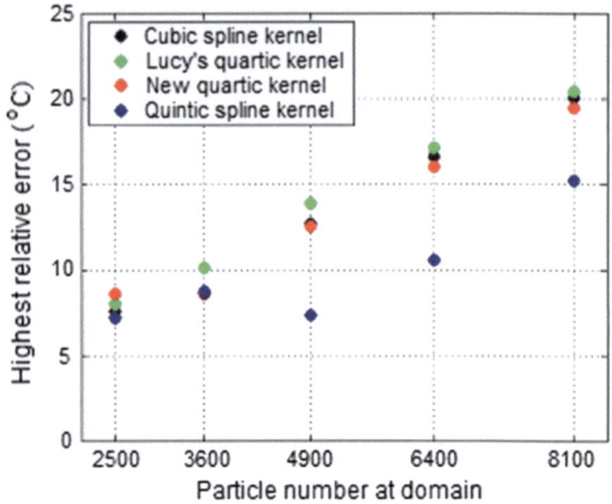

Fig. 4.7 Behaviour of the highest relative error

4.3 Still Liquid Inside an Immobile Reservoir

This is one of the first problems that must be understood and solved in the study of fluid mechanics. The fluid is Newtonian, incompressible, uniform and isothermal, and it is at rest within an immobile (and open to atmosphere) reservoir, at the same ambient temperature.

4.3.1 Physical-Mathematical Modelling

Analysing the fluid inside the reservoir from the physical conditions presented, we conclude:

- The conservation of mass is guaranteed for each Lagrangian element, because the fluid is isothermal and is at the same ambient temperature (there is no evaporation).
- The specific internal energy (a function of only the temperature for incompressible fluids) is constant, and the energy conservation equation must not be solved.
- The conservation of momentum equation (Newton's second law or Navier–Stokes equation for fluids) is the only equation to be solved in this problem.

The analytical solution is known and will be presented below.

The integration of the conservation of momentum equation (Eq. 2.2) in the vertical direction leads to Eq. (4.127):

$$(P - P_o) = \rho \text{g}(\text{H} - \text{y}) \tag{4.127}$$

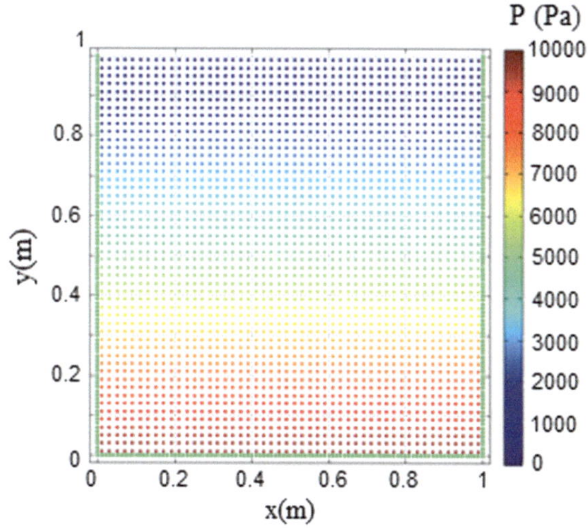

Fig. 4.8 Hydrostatic pressure field

where

P	is the absolute pressure
P_o	is the reference pressure (atmospheric pressure)
ρ	is the fluid density
H	is the ordinate of the free surface
y	is the ordinate of the fluid particle in study.

Figure 4.8 shows the pressure field on the Lagrangian particles at the domain.

In recent studies, standard SPH formulations have been used to solve the conservation of momentum equation (Eq. 2.2). In these studies, virtual/fictitious particles have been used in the boundary treatment. Repulsive models based on intermolecular forces (such as the Lennard-Jones force) have also been used in many researches today.

As discussed in Sect. 3.2.5, the approximations of the physical properties using the standard SPH method involve errors. This approach increases the difficulties in guaranteeing the balance of forces and the unchanging positions of particles over time, as shown in the results provided by Fourtakas et al. [8], Korzilius et al. [9], Vorobyev [10] and Goffin et al. [11].

Aiming to solve the problem of the oscillation of the particles, found in the previous literature results, a SPH modified modelling using the modified pressure concept (as found in [12]) is proposed in this book:

$$\boxed{P_{\text{mod}} = (P - P_o) - \rho g(\mathrm{H} - \mathrm{y})} \tag{8}$$

where P_{mod} is the modified pressure.

The use of the modified pressure concept changes the form of the conservation of momentum equation, as presented in Eq. (2.9):

$$\boxed{\rho \frac{D\vec{v}}{Dt} = -\nabla P_{\text{mod}} + \mu \nabla^2 \vec{v}} \quad (9)$$

Taking into account that $(P - P_o)$ is the pressure exerted by the fluid column on the particle with ordinate y, the modified pressure is zero on each particle in the domain in the hydrostatic problem. This implies that the modified pressure gradient on a fixed particle is also zero. Using the symmetric SPH form to the gradient, Eq. (3.32), we have:

$$\boxed{\left(\frac{\nabla P_{\text{mod}}}{\rho}\right)_i = -\sum_{j=1}^{n} m_j \left(\frac{P_{\text{mod}\,i}}{\rho_i^2} + \frac{P_{\text{mod}\,j}}{\rho_j^2}\right) \nabla W(X_i - X_j, h) = 0} \quad (4.128)$$

where

$P_{\text{mod}\,i}$ is the modified pressure on the fixed particle
$P_{\text{mod}\,j}$ is the modified pressure on the neighbouring particle
∇P_{mod} is the modified pressure gradient.

Since the modified pressure is zero on each particle in the domain, the gradient of the modified pressure (obtained from the Eq. (4.128)) is also zero.

The SPH equation of momentum conservation, written in an alternative way, was solved by adding the gradient of the modified pressure:

$$\boxed{\frac{D\vec{v}_i}{Dt} = -\left(\frac{\nabla P_{\text{mod}}}{\rho}\right)_i + 2\upsilon_i \sum_{j=1}^{n} \frac{m_j}{\rho_j}\left(\vec{v}_i - \vec{v}_j\right) \frac{(X_i - X_j)}{|X_i - X_j|^2} \cdot \nabla W(X_i - X_j, h)} \quad (4.129)$$

4.3.2 Numerical Simulations

Figure 4.9 shows the algorithm employed in numerical simulations. The geometry simulated is shown in Fig. 4.10.

The origin of the frame of reference, that has a positive vertical orientation pointing upwards, is located at the bottom of the reservoir. The analysis was performed in a 2-D domain. The dimensions of the transverse section are 1.0 m × 1.0 m.

In order to respect the classical continuum mechanical laws (in which the oscillation of the particles cannot exist and the molecular motion is not seen in hydrostatics problems), a solution using the modified pressure concept in conjunction with the reflective boundary conditions has been obtained. The reflective boundary treatment technique, based on Newton's law of restitution, fully respects the concepts of continuum mechanics.

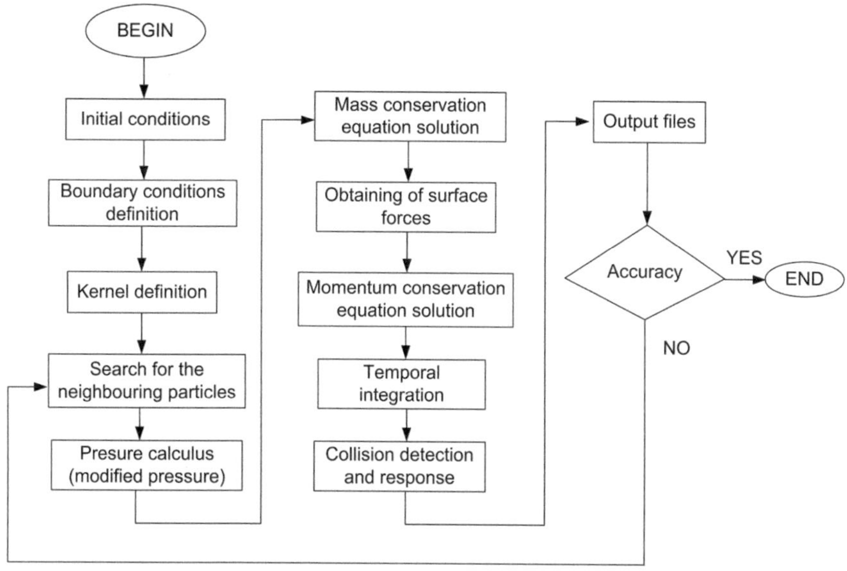

Fig. 4.9 Flowchart of the numerical code for the reservoir problem, employing the reflective boundary conditions

Fig. 4.10 Open reservoir containing a Newtonian, uniform, incompressible and isothermal fluid

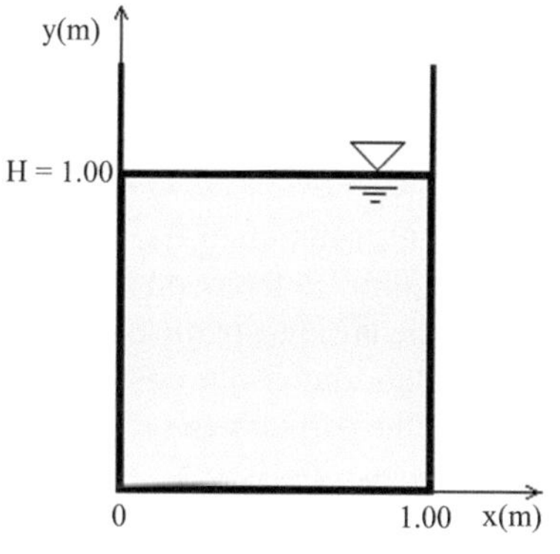

Two thousand five hundred particles were distributed regularly inside the domain (50 particles per side of the reservoir) at sea level and at 20 °C ($\rho_{water} = 1,000.00$ kg/m^3 and $\upsilon_{water} = 1.00 \times 10^{-6}$ m^2/s). The modified pressures and velocities of all particles have been initialized to zero. The time step of the simulations was 1.00×10^{-4} s. Corrections for the physical properties of the particles near the boundaries did not need to be performed. Cubic and quintic spline kernels have been used in the simulations. It has verified that the kernel choice does not influence the solution achieved.

Figure 4.11 shows the initial setup of the fluid particles, their centres of mass and the walls of the reservoir.

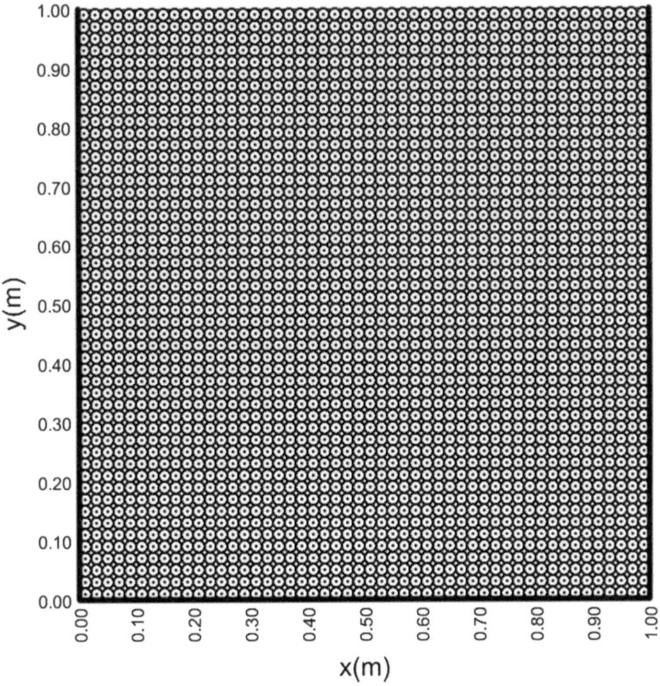

Fig. 4.11 Initial setup of particles and their centres of mass (represented by dots) and the solid boundaries

After the solution of Eq. (4.129) and updating the positions and velocities of the particles through temporal integration, an appropriate physical boundary treatment was applied using a particle collision detection and response algorithm.

The algorithm (in the collision detection stage) detected that the contacts occurred between the particles and the lateral and lower planes of the reservoir, even though the particles were at rest.

In the response stage, the reflections of the particles were carried out based on Eq. (3.116). In the contact points, r_p is equal to d (in which r_p is the particle radius and d is the distance between the centre of mass of the particles and the collision plane). The coefficients of restitution of kinetic energy and friction were zero, and the responses to the velocities of the particles were also zero. Consequently, the positions of the centres of mass were unchanged, maintaining the hydrostatic equilibrium. Thus, the field of modified pressures was renewed at each iteration with a null value, since the particles remained at rest (and free surface has not changed).

After starting the simulation, at the time instant 5.00 s (50,000 numerical iterations), the coincidence of the positions of the centres of mass at that moment and their initial positions was verified (regardless of the kernel used) in agreement with the analytical solution. The non-penetration between the particles and the absence of motion were guaranteed throughout the simulation time. Figure 4.12 shows the coin-

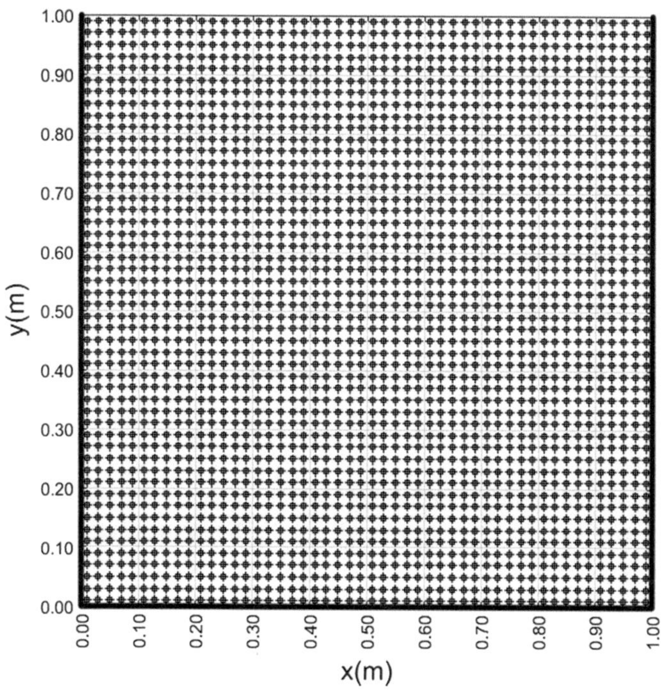

Fig. 4.12 Particles at rest inside the reservoir. The squares and the plus signs represent the centres of mass of the fluid particles at the time instant 0 and at the end of the simulation, respectively

cidence of the positions of the centres of mass at the final instant of the simulation (5.00 s/50,000 iterations) and their initial positions.

4.3.3 Comments

By making a brief analysis of the solution found, it is known that in the approximations of a function and its derivatives taken by the SPH method there are errors [13, 14]. When, in previous studies, the standard SPH method was applied without the use of modified pressure in order to obtain approximations for the pressure and viscosity forces (terms that were present in the momentum conservation equation and were added to the gravitational force) small mistakes appeared.

Despite the small orders of magnitude of those errors, they spread and increased during the numerical simulation. This resulted in the appearance of particles' motion. By using the modified pressure concept, an improved form of the momentum conservation equation has been achieved. The sum of the gravitational force and the terms approximated by the SPH for surface forces have been avoided, and consequently, the propagation of errors no longer occurred.

The numerical solution provided by the modified SPH method is in perfect agreement with the analytical solution and the continuum theory, with the complete exclu sion of any oscillatory movement seen in previous studies [8–11], in which fictitious or dynamic particles have been used in the boundary treatment.

The boundary conditions that employ fictitious (or virtual) particles and molecular forces should be avoided in continuum mechanics. The concepts of molecular modelling have no validity in the continuum domain, where the concepts and laws of classical mechanics are the ones that should be used. The reflective boundary condition technique is recommended because it respects the laws of continuum mechanics.

4.4 Dam Breaking Over a Dry Bed

The studies of dam breaks are of great importance in the prevention of accidents, which can cause serious environmental consequences, besides being a risk to the resident populations in the vicinity of those constructions.

Usually, in the standard SPH approach (a one-particle model), the collisions between the fluid particles in the flow are not considered. There are numerous studies on this subject in the literature. Monaghan [15] performed one of the first simulations with SPH for fluid flows. The repulsive boundary conditions (an analogy to Lennard-Jones intermolecular forces) were employed in these pioneering studies. Lee et al. [16] used virtual particles, types I and II, and applied the Neumann boundary condition for the pressures on the boundary particles. Korzilius et al. [9] have proposed an interparticle model for the SPH method (in which the collisions between the particles ensures that particles do not cluster) and a boundary treatment using ghost particles. Crespo et al. [17] simulated this problem employing dynamic particles in the boundary treatment achieving numerical results in agreement with the experiments.

4.4.1 Physical-Mathematical Modelling

The fluid inside the dam was considered incompressible, uniform and isothermal, and the mass conservation and Navier–Stokes equations, repeated below, were employed in simulations. The standard SPH approach (one-particle model) was implemented.

$$\boxed{\frac{D\rho_i}{\mathrm{Dt}} = \sum_{j=1}^{n} m_j \left(\vec{\mathrm{v}}_i - \vec{\mathrm{v}}_j\right) \cdot \nabla W(X_i - X_j, h)} \tag{37}$$

$$\begin{aligned} \frac{D\vec{\mathrm{v}}_i}{Dt} &= -\sum_{j=1}^{n} m_j \left[\frac{P_i}{(\rho_i)^2} + \frac{P_j}{(\rho_j)^2}\right] \nabla W(X_i - X_j, h) + \\ & 2\upsilon_i \sum_{j=1}^{n} \frac{m_j}{\rho_j} \left(\vec{\mathrm{v}}_i - \vec{\mathrm{v}}_j\right) \frac{(X_i - X_j)}{\left|X_i - X_j\right|^2} \cdot \nabla W(X_i - X_j, h) + \vec{g} \end{aligned} \tag{38}$$

The absolute pressure acting on the particles presents terms related to the hydrostatic and dynamic pressures. The prediction of this last parcel was made using the Tait equation.

4.4.2 Numerical Simulations

The reservoir (at sea level at 20 °C had a length of 0.420 m, a height of 0.440 m and a depth of 0.228 m). Numerical simulations were performed in a 2-D computational domain: a rectangular area with a height of 0.228 m and a width of 0.114 m (aspect ratio of 2). The dammed volume of water was discretized by 36 columns containing 71 fluid particles each (totalling 2,556 fluid particles with a lateral distance between their centres of mass of 3.257 mm).

Figure 4.13 brings a flowchart of the numerical code employed in the simulations of dam breaking.

Figure 4.14 shows the domain simulated and the initial particle setup.

In the simulations, which used the reflective boundary conditions, the particles of the free surface were initially marked and the pressures on them were initialized to zero. It is known that the pressure on the free surface is the atmospheric [18]. This pressure was defined as the reference zero pressure.

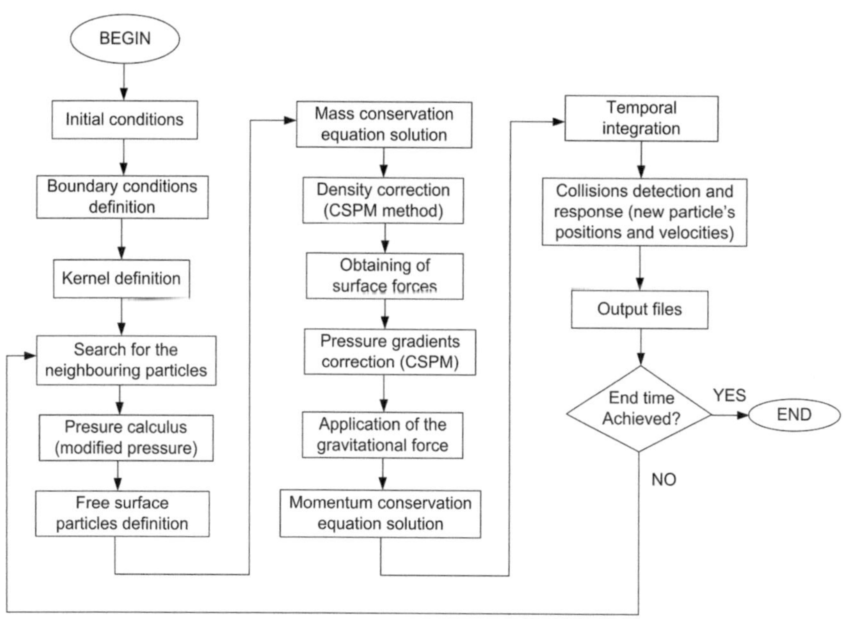

Fig. 4.13 Flowchart of the algorithm for dam breaking simulation

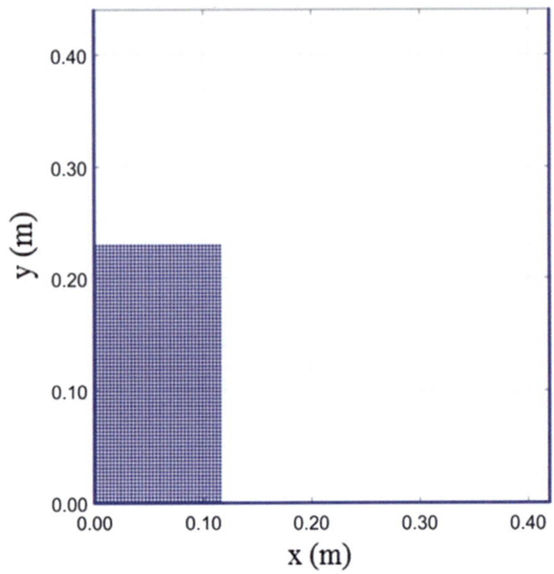

Fig. 4.14 Simulated 2-D computational domain and the initial distribution of the centres of mass of the particles

The particles on the free surface must continue in this condition until the wave impacts the wall of the reservoir. For this reason, the pressure on the marked particles was renewed at every numerical iteration (zero constant pressure condition).

The free surface has been monitored. It was seen that the marked particles continued to define the free surface during the simulation time. Thus, it has verified that the reflective boundary conditions (applied to the walls of the reservoir) respected the Newman boundary conditions (applied at the free surface of the fluid), not changing the expected shape of the moving wave.

To avoid numerical instabilities and interpenetration between particles, the artificial viscosity was added to the momentum conservation equation, as shown in Eq. (3.69). The coefficient α_{π} received the values of 0.20 and 0.30, while the coefficient β_{π} received the value of 0.00. No turbulence model was used in the simulations.

Cubic spline kernel has been used in the simulations. Simulations were performed for a physical time of 0.30 s (30,000 numerical iterations). Corrections of the physical properties (density and pressure gradient) of the particles located near the boundaries were made every 30 numerical iterations using a correction factor based on the Corrective Smoothed Particle Method (CSPM). XSPH correction has not been used in simulations. The time step used was 1.0×10^{-5} s.

The term B in the Tait equation received the value of 0.85×10^{5} Pa, and γ was equal to 7. The total pressure acting on a water particle was the sum of two parcels: one dynamic and the other exerted by the water column above the reference particle.

After obtaining the accelerations of the particles from the solution of the momentum conservation equation by SPH method and realization of the temporal integration

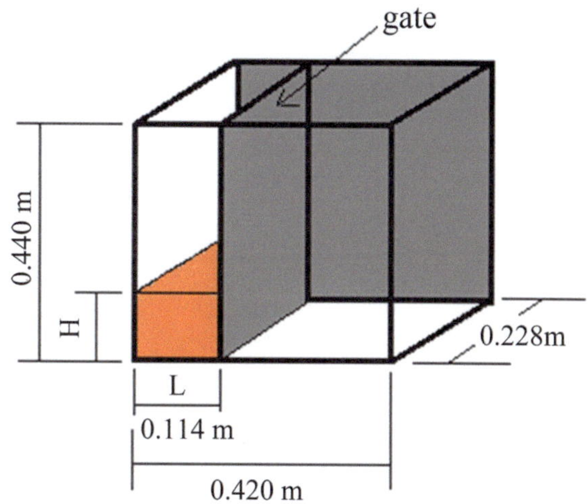

Fig. 4.15 Schematic diagram of the experimental apparatus.

using Euler's method, the position of the centre of mass of each fluid particle was obtained. The collision and response algorithm returned the particles colliding with the planes into the domain. The degree of elasticity of the particles' collisions against the reservoir wall was measured by a coefficient of restitution of kinetic energy. The collisions have been considered perfectly elastic (CR = 1.00). The coefficient of friction was 0.00.

The wave front position was monitored before its collision with the right-hand wall of the tank. Aiming to validate the results obtained from the collision detection and response algorithm, the experimental data provided by Cruchaga et al. [19] have been used. The experimental apparatus used by the researchers is shown schematically in Fig. 4.15.

The percentage differences between the abscissas of the wave fronts obtained from SPH and experimentally were calculated from the expression:

$$\Delta x_p = \frac{\left(x_{exp} - x_{SPH}\right)}{x_{exp}} 100 \tag{4.130}$$

where

x_{exp} and x_{SPH} are the abscissas of the wave fronts obtained from experiments and SPH simulations, respectively

Δx_p is the percentage difference between the abscissas of the wave fronts.

The greatest differences between the wave fronts (experimental and provided by the SPH/collision algorithm) occurred in the initial instants of the dam breaking. At 0.10 s, the difference was 6.90% (0.2032 m SPH/0.2170 m experimental). As time progressed, the difference decreased, and at 0.20 s, it was 4.40% (0.3776 m SPH/0.3610 m experimental).

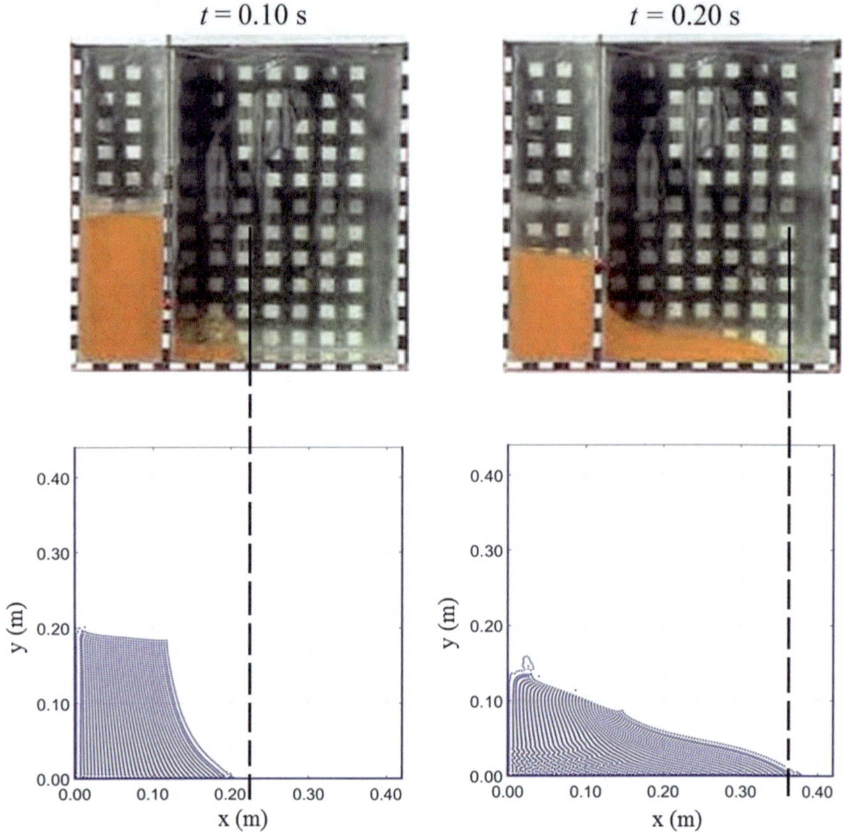

Fig. 4.16 Comparison between the experimental and numerical results

Figure 4.16 shows a graphical illustration of the experimental (provided by [19]) and numerical results.

Considering that the solution provided by the SPH method for the fluid physical properties, which includes the approximations of the positions and velocities of the particles (that are the input data for the collision and detection algorithm) present errors [13, 14], the numerical results showed good agreement with the experimental results within the simulation period.

4.4.3 Comments

At the end of this case study, it was possible to verify the applicability of the reflective boundary conditions in dynamic problems, the restoration of the consistency in the regions close to the contours (realized through the CSPM method) and the effect of the artificial viscosity in the solution convergence.

The differences among the numerical and experimental results are due to errors arising from the solution provided by the Lagrangian particle method [20].

The use of the reflective boundary conditions respected the laws of the continuum.

4.5 Oil Spreading on a Calm Sea

When oil is spilled in the ocean, it initially spreads on the water surface. Oil spill can occur during the various stages of well drilling and repair operations, when the oil is being produced from operations at sea, handled and temporarily stored, or when it is being transported at sea, through flow lines, submarines, or by ship.

The spreading is due to the pollutant's tendency to flow over itself. The knowledge of the physical properties related to this transportation process of the pollutant, such as velocities and positions, is of fundamental importance for the adoption of timely measures to protect the environment.

4.5.1 Motivation

Oil spills cause large and negative impacts on coastal ecosystems such as mangroves, marshes, beaches, rocky shores, tidal plains, coral reefs and sandy/rocky bottoms, which exhibit high biological richness and trophic complexity, interacting with each other through transfer of energy, nutrients, migrations and the reproductive cycle of species. With oil spreading over sea water, the spots can reach, for example, the intertidal zone of mangroves, causing severe impacts. The organisms that live on the surface of the water are the first ones to be reached (plankton, in the base of the trophic chain). When they hit the coast, oil waste interacts with sediments such as beach sand and gravel, stones and rocks, vegetation and terrestrial habitats of wild animals and humans on the path of drift wildlife, causing erosion and contamination [21].

The preservation of coastal environments in all its aspects—fauna and flora, social, economic, public health, leisure, among others—is an environmental concern. The study of the spreading of oil aims to obtain the oil slick reach forecasting over time. By acting in the most immediate way possible, it is possible the accomplishment of containment operations for the protection of the fauna and flora of the regions close to the spill.

The spreading is the most significant transport process in the first few hours of the spill. Its understanding and quantification are of great importance to the preservation of the environment.

Particle methods, which address the spreading problem with a Lagrangian view for the two phases involved (oil and water), are incipient in the scientific study of spreading. Violeau [22], and Yang and Liu [23], presented their model and results employing this approach in the study of oil spill containment using floating devices.

The phenomenon is divided into three distinct stages, in which the gravitational, inertial and viscous forces and superficial tension are held accountable for the movement of the oil over the water. On a calm sea, the shear stress due to the wind action on the surface of the oil, the advective forces and the turbulent diffusion are not considered. Due to the inexistence of winds, currents and waves, the oil spreads in a circular shape, in a displacement that presented radial symmetry.

Fay [24,25] adjusted curves to the experimental data and obtained expressions used in prediction models for the oil slick reaching. These adjusted curves defined for idealized theoretical conditions are still used, with some modifications.

It is known that the oil spreading, in real-life conditions, cannot be fully explained by these equations; however, through modifications, like considering the shear effects caused by the wind on the surface of the oil (which explain the elliptical shape of the oil slick in an actual spreading), the equations can be applied nowadays.

Currently, most of the models for the prediction of oil slick are hybrids and utilize meshes (in the Eulerian modelling of the water) and particles in the discretization of the oil. The input data for the simulation of the gravity-viscous stage, the longest and the one that causes the greatest environmental impacts, it is in general, obtained from a prediction based on Fay's equation. The objective of this case study is to implement and validate a purely Lagrangian model, capable to provide the input data for the simulation of the second phase of oil spreading (gravity-viscous) without using the curve adjusted by Fay for the oil slick prediction.

This case study is directed towards the understanding of the first stage of the spreading (gravity-inertial regime), disregarding the other processes (physical, chemical and biological) that occur during the spill, such as evaporation, vertical dispersion, emulsification, photo-oxidation, dissolution, biodegradation and sedimentation, schematically illustrated in Fig. 4.17.

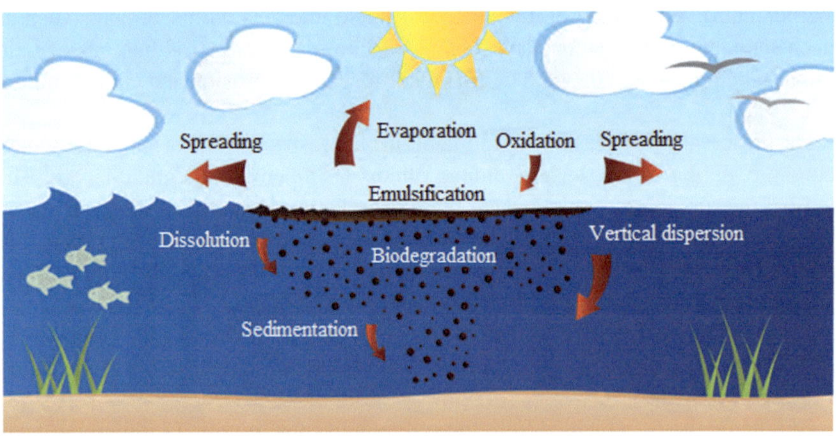

Fig. 4.17 Main physical, chemical and biological processes occurring at an oil slick in the sea

4.5.2 Physical-Mathematical Modelling

The SPH method was used in discretization of the equations of conservation of mass and momentum of oil, in a calm sea condition.

In the first stage of the spreading, gravity predominates as a driving force and inertia is the main force that is resistant to the oil motion [24,25]. This stage has its duration as a function of the volume of the spilled oil. The expression adjusted by Fay to the experimental results is shown in Eq. (4.131):

$$\boxed{\mathrm{D} = 2k_1\left(\Delta_{\mathrm{w}} g V t^2\right)^{\frac{1}{4}}} \tag{4.131}$$

$$\Delta_{\mathrm{w}} = \frac{\rho_{\mathrm{w}} - \rho_o}{\rho_{\mathrm{w}}} \tag{4.132}$$

where

D is the oil slick diameter
t is the time
ρ_{w} is the water density
ρ_o is the oil density
V is the volume of the oil spilled $k_1 = 1.14$.

This stage of the spreading lasts until the time instant is t_f given by Eq. (4.133):

$$\boxed{t_f = \left(\frac{\mathrm{h}_o}{g\Delta_{\mathrm{w}}}\right)^{\frac{1}{2}}} \tag{4.133}$$

where $\mathrm{h_o}$ is the initial oil height.

According to Stolzenbach et al. [26], the initial movement of the spilled oil is analogous to that of a dam break. This consideration was taken into account, and the

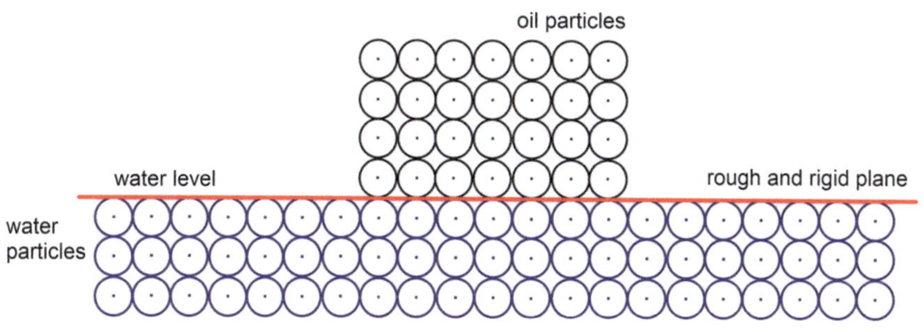

Fig. 4.18 Oil and the water particles separated by a rough and rigid plane

mass and momentum conservation equations have been solved by the SPH method (Eqs. (3.37) and (3.38) were applied for the oil phase).

In the interface treatment, the interaction between the phases (water and oil) was measured from the definition of a horizontal, rough and rigid plane fixed over the water level as shown in Fig. 4.18, over which the oil spreads. The non-penetration condition and reflections of the oil particles when colliding against the interface plane have been implemented.

Artificial repulsive forces have not used in the interface treatment, because they mix concepts of molecular microscopic dynamics and continuum mechanics. The water was considered at rest during the spreading (horizontal dispersion) of the oil. The roughness of the plane fixed on its surface promoted the equivalent effect of slowing down the movement of the oil (realistically performed by water at the liquid interface). The interaction between oil and water at the interface has been modelled by considering the elasticity and friction involved in the collisions of the oil particles against the plane. An algorithm was used aiming to detect the collisions of the oil particles against the plane surface and to provide a physical response.

The water particles (below the rough and rigid plane) were maintained at rest (null velocities), and the pressure acting on them was defined as zero (invariant). The static (due to the oil column over the reference particle) and dynamic pressures acted on the oil particles. The Tait equation was employed to predict the dynamic pressure field of the oil. The pressures on the oil particles in contact with the atmosphere were started and maintained with 0.00 Pa throughout the simulations.

4.5.3 Numerical Simulations

The simulated domain consisted of a tank with a 20-m length, 20-m width, 1.5-m height, a water level of 0.50 m and an oil volume initially disposed, in the centre of the tank, in the form of a cylinder with a height and diameter of 1 m. Figure 4.19 presents a schematic 3-D illustration of the oil cylinder over the water surface at the initial instant of the spill.

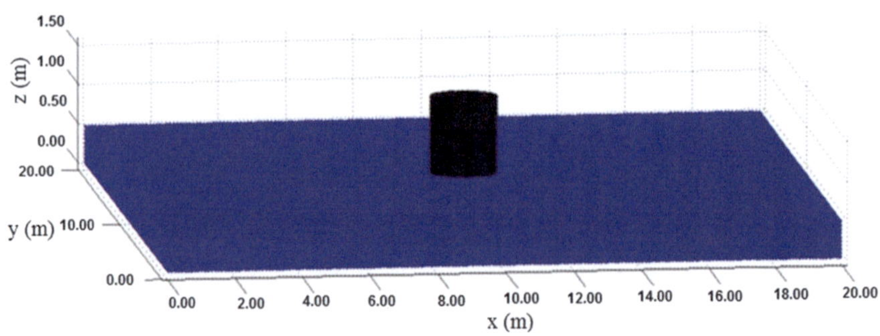

Fig. 4.19 Schematic 3-D illustration of the oil cylinder over the water surface at the initial instant of the spill

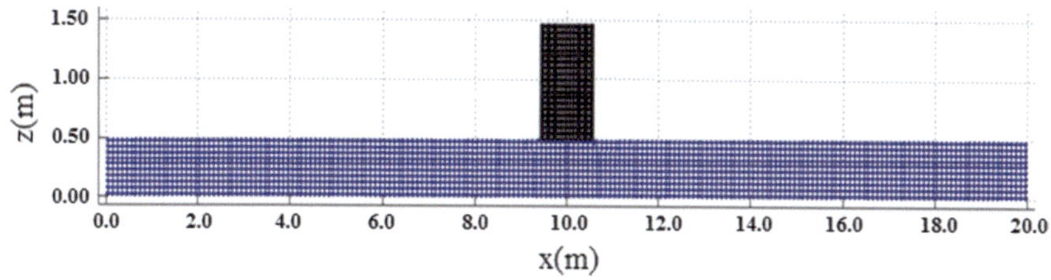

Fig. 4.20 Longitudinal section of the oil cylinder at the initial time instant

Fig. 4.21 Upper view: the circular shape of the oil slick spreading on a calm sea condition

A two-dimensional (axisymmetric) analysis of the problem was carried out. A longitudinal section of the volume was studied, as shown in Fig. 4.20. Figure 4.21 shows an upper view of the spreading and circular shape of the oil slick spreading on a calm sea condition.

The study was performed with a small volume of oil, which caused a very short first stage of spreading. A higher volume of oil was used in Fay's experiments; however, there were no restrictions in his work regarding the volume of the oil spill and the use of his adjusted curves in the prediction of the slick diameter, which made the simulations feasible, in relation to the number of particles used in the discretization of the domain.

The positions of the oil particles along the horizontal direction were monitored at each numerical iteration. The initial lateral distances between the centres of mass of the particles (equal to the particle diameter) were 2.00×10^{-2} m. The radius of the domain of influence was equal to 5.00×10^{-2} m. A total of 2,500 oil particles and 26,000 water particles (density of 1000.00 kg/m^3 and absolute viscosity of 1.00×10^{-3} Pa s) were employed in the discretization of the domain. The oil was considered a homogeneous, uniform and isotropic fluid. The simulations were carried out for lightweight oil, with a density of 850.00 kg/m^3 and absolute viscosity of 3.32×10^{-3} Pa s. The duration of the spreading in its first stage, obtained by Eq. (4.133), was 0.82 s with a total of 5,815 numerical iterations.

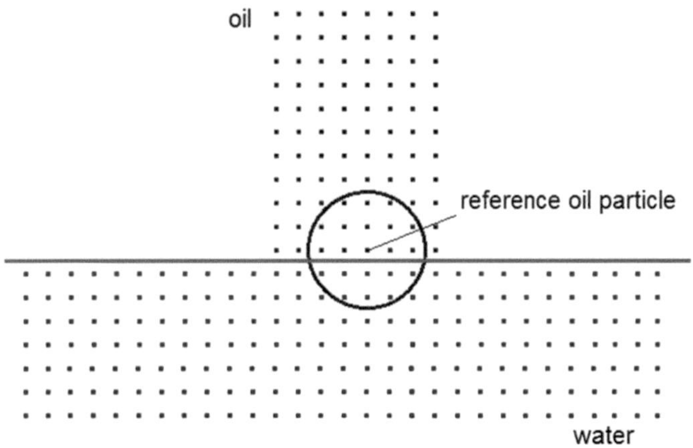

Fig. 4.22 Domain of influence of an oil particle near the interface. The water particles neighbouring the reference oil particle were identified (All particles are represented by their centres of mass)

Fig. 4.23 Evolution of the reach of the oil slick in time. The inner circumference refers to the initial diameter of the oil (at the beginning of the spill)

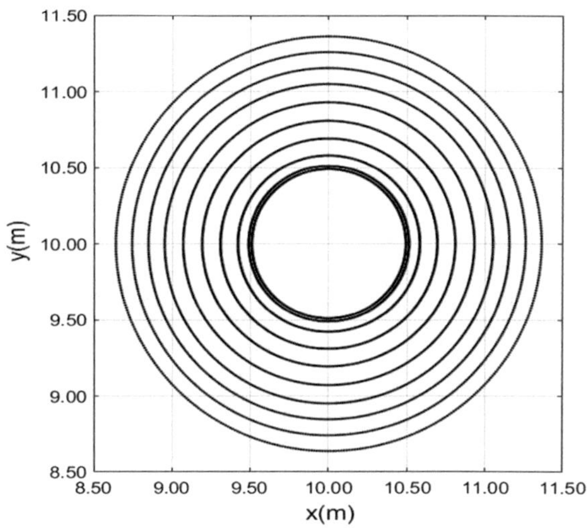

Figure 4.22 presents the centres of mass of the particles and the domain of influence for a reference oil particle. In the SPH particle method used in the simulations, there was the contribution of the neighbouring particles on the oil properties. Thus, the physical properties of the water (pressures, densities, viscosities, velocities and others) influenced the physical properties of the oil spreading near the interface.

After obtaining the velocity fields of the oil particles using the SPH method, the collisions that occurred against the horizontal plane established at water level were detected with the analysis of the oil particle trajectories. Once a collision against the plane was detected, the particle was reflected, considering the energy losses (in the particle shocks and friction).

The evolution of the oil particles on the water surface was monitored during the simulations. The circumferences that defined the spreading range, in each numerical iteration, were defined with a radius equal to the highest distance achieved by a particle (represented by its centre of mass), measured from the centre of the tank (10.00, 10.00, 0.50 m). In Fig. 4.23, the plotted circumferences show the evolution of the reach of the oil slick in time.

The calibration of the physical-mathematical model was performed through tests that used different coefficients of restitution of kinetic energy and friction, aiming to reach the results obtained by the curve adjusted by Fay to the experimental data.

The parameters used for the simulations performed are stated below:

- The direct search for neighbouring particles has been employed.
- The density renormalization of the oil has been performed (CSPM method).
- In the prediction of dynamic pressure field on the particles, the parameter B received the value 0.85×10^5 Pa.
- The particle radius, used in the verification of the collisions against the plane, was equal to 1.00×10^{-2} m.
- The pressure gradients have been corrected using the CSPM method.
- The SPH interpolations have been carried out with the cubic spline kernel.
- The smoothing length varied over time and was updated with the aid of Eq. (3.62).
- The artificial viscosity, a numerical correction widely used in SPH simulations, was not used.

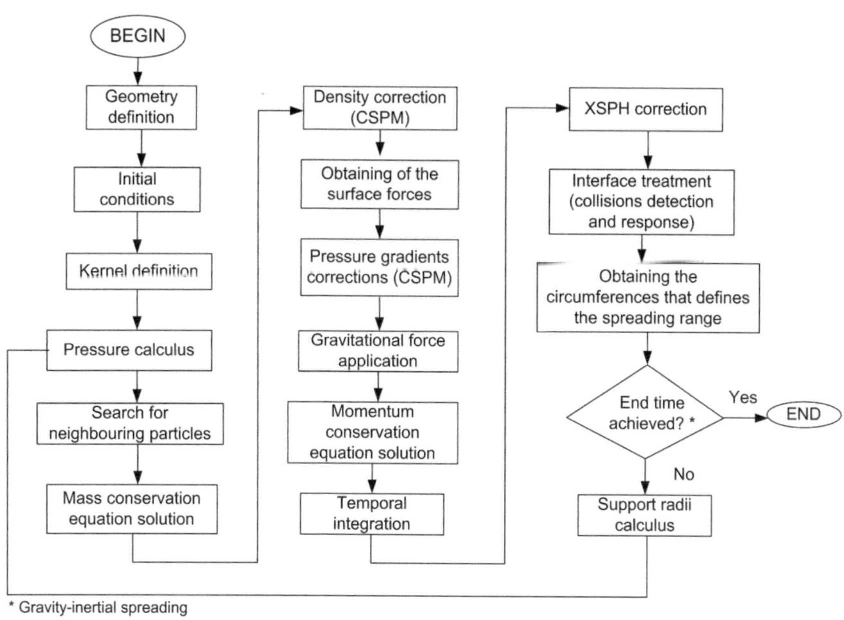

Fig. 4.24 Flowchart of the algorithm used in the oil spreading simulations

- No turbulence model was employed.
- The particles were moved at an average velocity of their interaction partners: the XSPH method was applied ($\eta = 0.50$).
- The Runge–Kutta first-order integration method (Euler's method) was used to update the positions and velocities of the particles.
- The time step employed in all simulations, defined by the Courant–Friedrichs–Lewy (CFL) stability criterion, was 1.41×10^{-4} s.

The flowchart of the algorithm used in simulations is presented in Fig. 4.24.

4.5.4 Results and Discussions

Simulations were performed to calibrate the physical-mathematical model of dam breaking of the oil cylinder over the water surface. Numerical tests were made using different coefficients of restitution of kinetic energy and friction. The validation of the numerical results was performed through the comparisons with the results provided by Fay's equation at the end of the gravity-inertial spreading ($t_f = 0.82$ s).

SPH simulations were performed for CR receiving values from 0.85 to 0.92 and CF ranging from 0.15 to 0.10. The oil slick diameter predicted by Fay's equation, Eq. (4.131), was 3.099 m. The results including the differences between the predicted oil slick diameters (ΔD = Fay's result − SPH) and the per cent error are presented in the last two columns of Table 4.4.

The evolution of the oil slick from the beginning of the spill to the end of the gravity-inertial spreading, for different combinations of friction and kinetic energy restitution coefficients, is shown in Figs. 4.25 and 4.26. The upper curves present Fay's solution, and the lower curves present the SPH solutions.

It was verified that the oil slick prediction obtained by the SPH method converged to the results provided by Fay's equation at the end of the gravity-inertial spreading, when the coefficient of friction obtained the value of 0.10 and the per cent error was

Table 4.4 Oil slick diameter at the end of the gravity-inertial stage

CR	CF	Oil slick diameter prediction (m)		Difference between results – ΔD (m)	Per cent error[b]
		SPH method	Fay's result		
0.85	0.15	2.908		0.191	6.16
0.85	0.10	3.067		0.032	1.03
0.90	0.15	2.915	3.099[a]	0.184	5.94
0.90	0.10	3.071		0.029	0.94
0.92	0.15	2.933		0.166	5.36
0.92	0.10	3.126		−0.027	−0.90

[a]Predicted by Eq. (4.131)
[b]Calculated by expression (ΔD/Fay's result) × 100

0.90%. For CR equal to 0.90, the per cent error was 0.94, and for CR equal to 0.92, it was 0.90% (the better result).

From the result analysis, it can be stated that the dam breaking model applied in the oil spreading simulation (hypothesis of Stolzenbach and collaborators [26]) was adequate. The calibration of the model of collisions through the correct choice of the coefficients CR and CF was also adequate and important when obtaining the Lagrangian solution.

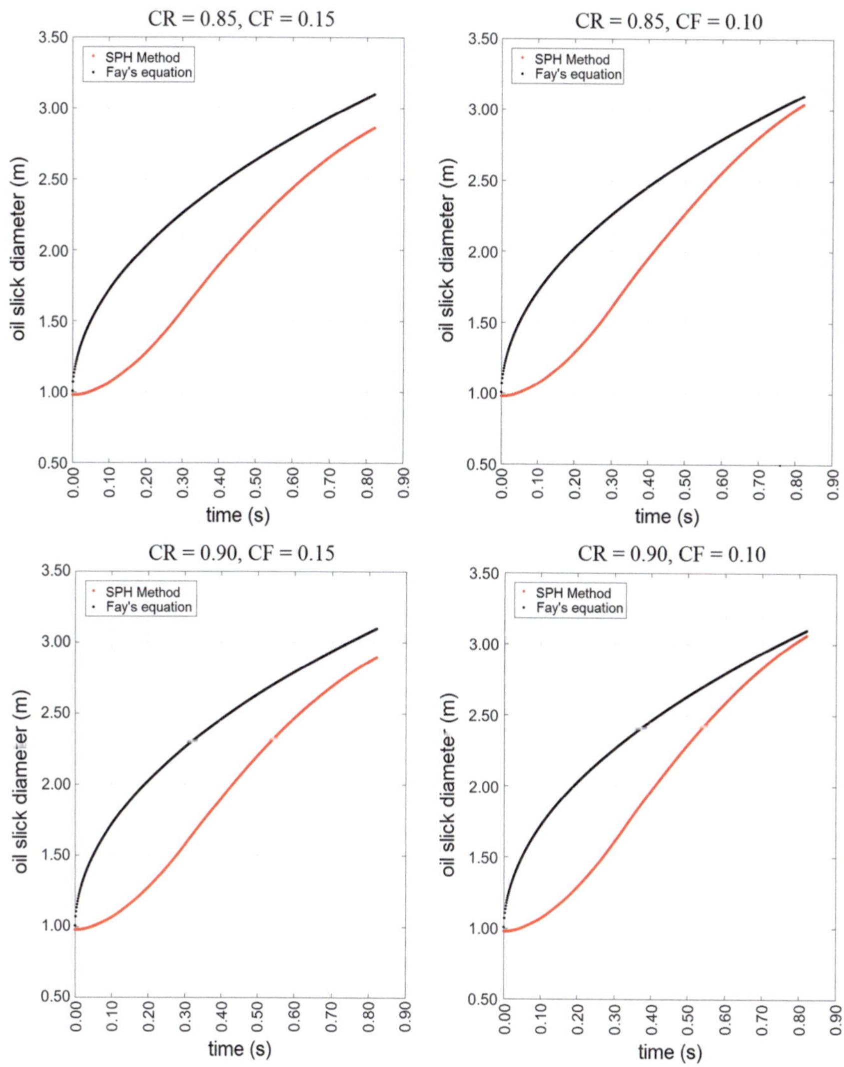

Fig. 4.25 Simulations results for CR 0.85 and 0.90 and CF equal to 0.15 and 0.10

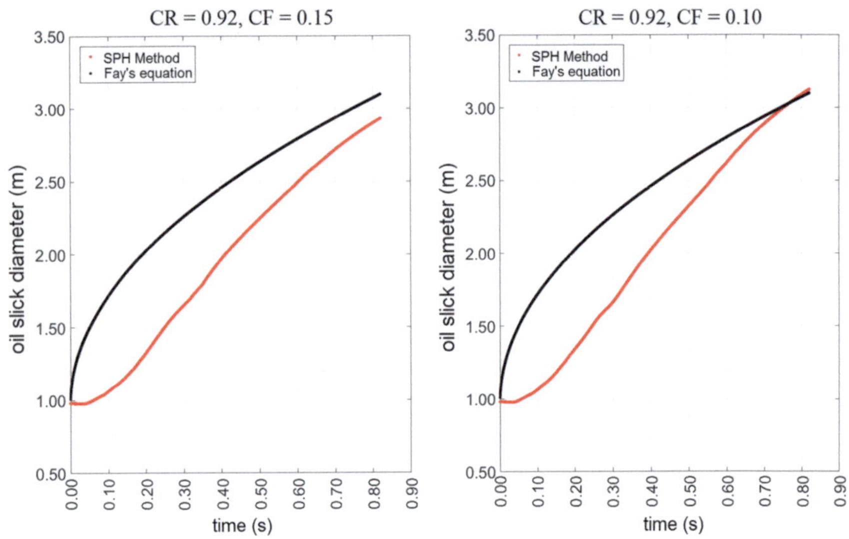

Fig. 4.26 Simulations results for CR 0.92 and CF equal to 0.15 and 0.10

For other types of oil, with specific physical properties (density, viscosity), as well as different spilled volumes, another value for the coefficients of restitution and friction can be found, so that there is an agreement between the Lagrangian results and Fay's solution in the end of the gravity-inertial stage.

The numerical result achieved is relevant because it can be used as the input data for the simulation of the subsequent stage of the oil spreading (gravity-viscous), the stage that causes the greatest environmental impact. One does not need to employ Fay's adjusted curve for its prediction anymore.

The excellent result obtained points to the continuity of the model development. It can be applied to real conditions in which the spreading occurs (where winds, currents and waves are present), by assigning the physical properties of these external agents to the particles that discretize the pollutant.

4.6 Concluding Remarks

This chapter presented the application of Smoothed Particle Hydrodynamics in the solution of some problems in continuum mechanics.

Final comments will be made on them:

1. In the first problem, the boundary effect was seen, and the kernel truncation, in the temperature distribution obtained, as well as the influence of the kernel and particle numbers in the solution found. An improvement in the results can be

achieved by a correction of the Laplacian of temperature in the regions where the kernel truncation occurred.

2. In the hydrostatic problem, the next case studied, the problem of the particle oscillation has been solved. The SPH results are in agreement with the analytical solution. It is recommended to use reflective boundary conditions for it to be a totally physical approach and to respect the continuum hypothesis.
3. The dam breaking problem has been solved using the reflective boundary conditions. The numerical results has been satisfactory: the divergence between the SPH and experimental results was 4.40%. Other higher-order temporal integration methods can be employed in order to verify the improvement of the accuracy of the input data for the collision and response algorithm [20,27]. Thus, more accurate numerical results can be achieved.
4. From the analysis of the hydrostatic and hydrodynamic results, it is concluded that the application of the reflective boundary conditions was adequate in both cases.
5. The expected diameter for the oil slick was obtained in the end of the first stage of the spreading using a purely Lagrangian model. Thus, it will not be necessary to use the Fay equation to provide this data for the subsequent phase (gravity-viscous) simulation.

References

1. Maruyama, S.: Molecular dynamics method for microscale heat transfer. In: Minkowycz, W.J., Sparrow E.M. (eds.) Advances in Numerical Heat Transfer, vol. 2, Chap. 6, pp. 189–226. Taylor & Francis, New York (2000)
2. Wang, X., Xu, X.: Molecular dynamics simulation of heat transfer and phase change during laser material interaction. J. Heat Transf. **124**, 265–274 (2002)
3. Pletcher, R.H., Tanehill J.C., Anderson, D.A.: Computational Fluid Mechanics and Heat Transfer, 3rd edn. CRC Press (2013)
4. Dilts, G.A.: Moving-least-squares-particle hydrodynamics-I. Consistency and stability. Int. J. Numer. Methods Eng. **44**, 1115–1155 (1999)
5. Stranex, T., Wheaton, S.: A new corrective scheme for SPH. Comput. Methods Appl. Mech. Eng. **200**, 392–402 (2011)
6. Korzilius, S.P., Schilders, W.H.A., Anthonissen, M.J.H.: An improved CSPM approach for accurate second-derivative approximations with SPH. J. Appl. Math. Phys. **5**(1), 168–184 (2017)
7. Belytschko, T., Krongauz, Y., Dolbow, J., Gerlach, C.: On the completeness of meshfree particle methods. Int. J. Numer. Methods Eng. **43**, 785–819 (1998)
8. Fourtakas, G., Vacondio, R., Rogers, B.D.: On the approximate zeroth and first-order consistency in the presence of 2-D irregular boundaries in SPH obtained by the virtual boundary particle methods. Int. J. Numer. Methods Fluids **78**(8), 475–501 (2015)
9. Korzilius S.P., Kruisbrink, A.C.H., Schilders, W.H.A., Anthonissen, M.J.H., Yue, T.: Momentum conserving methods that reduce particle clustering in SPH. CASA-Report 2014–2015, Eindhoven: Technische Universiteit Eindhoven. https://pure.tue.nl/ws/files/3858217/376670351851652.pdf (2014). Accessed 30 Apr 2017
10. Vorobyev, A.: A Smoothed Particle Hydrodynamics Method for the Simulation of Centralized Sloshing Experiments. KIT Scientific Publishing, Germany (2013)

11. Goffin, L., Erpicum, S., Dewals, B.J., Pirotton, M., Archambeau, P.: Validation of a SPH model for free surface flows. In: Proceedings of the 3rd SimHYDRO Conference, pp. 11–13, Nice, France (2014)
12. Batchelor, G.K.: An Introduction to Fluid Dynamics, 3rd edn. Cambridge University Press, UK (2000)
13. Liu, M.B., Liu, G.R.: Smoothed particle hydrodynamics (SPH): an overview and recent developments. Arch. Comput. Methods Eng. **17**, 25–76 (2010)
14. Vaughan, G.L., Healy, T.R., Bryan, K.R., Sneyd, A.D., Gorman, R.M.: Completeness, conservation and error in SPH for fluids. Int. J. Numer. Methods Fluids **56**, 37–62 (2008)
15. Monaghan, J.J.: Simulating free surface flows with SPH. J. Comput. Phys. **110**, 399–406 (1994)
16. Lee, E.-S., Moulinec, C., Xu, R., Violeau, D., Laurence, D., Stansby, P.: Comparisons of weakly compressible and truly incompressible algorithms for the SPH mesh free particle method. J. Comput. Phys. **227**(18), 8417–8436 (2008)
17. Crespo, A.J.C., Dominguez, J.M., Barreiro, A., Gómez-Gesteira, M., Rogers, B.D.: GPUs, a new tool of acceleration in CFD: efficiency and reliability on smoothed particle hydrodynamics methods. PLoS ONE **6**(6) (2011). https://doi.org/10.1371/journal.pone.0020685
18. Newman, J.N.: Marine Hydrodynamics. The Massachusetts Institute of Technology Press, Massachussets, USA (1977)
19. Cruchaga, M.A., Celentano, D.J., Tezduyar, T.E.: Collapse of a liquid column: numerical simulation and experimental validation. Comput. Mech. **39**, 453–476 (2007)
20. Fraga Filho, C.A.D.: An algorithmic implementation of physical reflective boundary conditions in particle methods: collision detection and response. Phys. Fluids **29**, 113602 (2017). https://doi.org/10.1063/1.4997054
21. Water Encyclopedia: Oil spills: impact on the ocean. http://www.waterencyclopedia.com/Oc-Po/Oil-Spills-Impact-on-the-Ocean.html. Accessed 03 Feb 2018
22. Violeau, D., Buvat, C., Abed-Meraim, K.: Numerical modelling of boom and oil spill with SPH. Coast. Eng. **12**, 895–913 (2007)
23. Yang, X., Liu, M.: Numerical modeling of oil spill containment by boom using SPH. Phys. Mech. Astron. **56**(2), 315–321 (2013)
24. Fay, J.A.: The spread of oil slicks on a calm sea. In: Oil on the Sea, pp. 53–64. Plenum Press (1969)
25. Fay, J.A.: Physical processes in the spread of oil on a water surface. In: International Oil Spill Conference Proceedings, pp. 463–467 (1971)
26. Stolzenbach, K.D., Madsen, O.S., Adams, E.E., Pollack, A.M., Cooper, C.K.: A Review and Evaluation of Basic Techniques for Predicting the Behavior of Surface Oil Slicks. Massachusetts Institute of Technology (1977)
27. House, D.H., Keyser, J.C.: Foundations of Physically Based Modeling and Animation. CRC Press, Taylor & Francis Group, Boca Raton, Florida, USA (2017)

Computer Code

Below follows the presentation of the general SPH computer code, developed for fluid dynamics and transport phenomena simulations, and its routines.

1. The initial positions, velocities, densities, temperatures, support radius and physical properties of the fluid particles are set to the beginning of the simulation. Besides, the boundary condition technique is chosen and the problem geometry is defined.
2. The smoothing function used in SPH interpolations is chosen.
3. Search for neighbouring particles: particles within the domain of influence of a reference particle may vary over time, and the search must be performed at each numerical iteration.
4. Pressure field prediction: in this routine, the prediction of the pressure field acting on the particles is performed. The absolute pressure has two parcels: the hydrostatic pressure (due to the fluid column over the point of the flow considered) and the dynamic pressure (predicted by the equation of state).
5. Mass conservation equation solution: the calculus of the density of each particle is provided by the SPH method.
6. Obtaining the surface forces: approximations to pressure and viscosity forces acting on the particles are obtained.
7. External forces application: the last term on the right side of Eq. (3.38) is the sum of all external forces per unity of mass. In the problems presented in this book, the only external force is gravity. Other forces are considered external forces in the literature such as the repulsive exerted by the virtual particles of type I on the fluid particles (Lennard-Jones molecular force) and free surface forces, if they are employed in the studied problem. However, there is the aforementioned conceptual contradiction of applying molecular concepts in the continuum domain (Sect. 2.1).
8. Momentum balance equation solution: the field of the accelerations of the particles is obtained.

C. A. Dutra Fraga Filho, *Smoothed Particle Hydrodynamics*,
https://doi.org/10.1007/978-3-030-00773-7

Fig. 1 Flowchart of a general SPH code

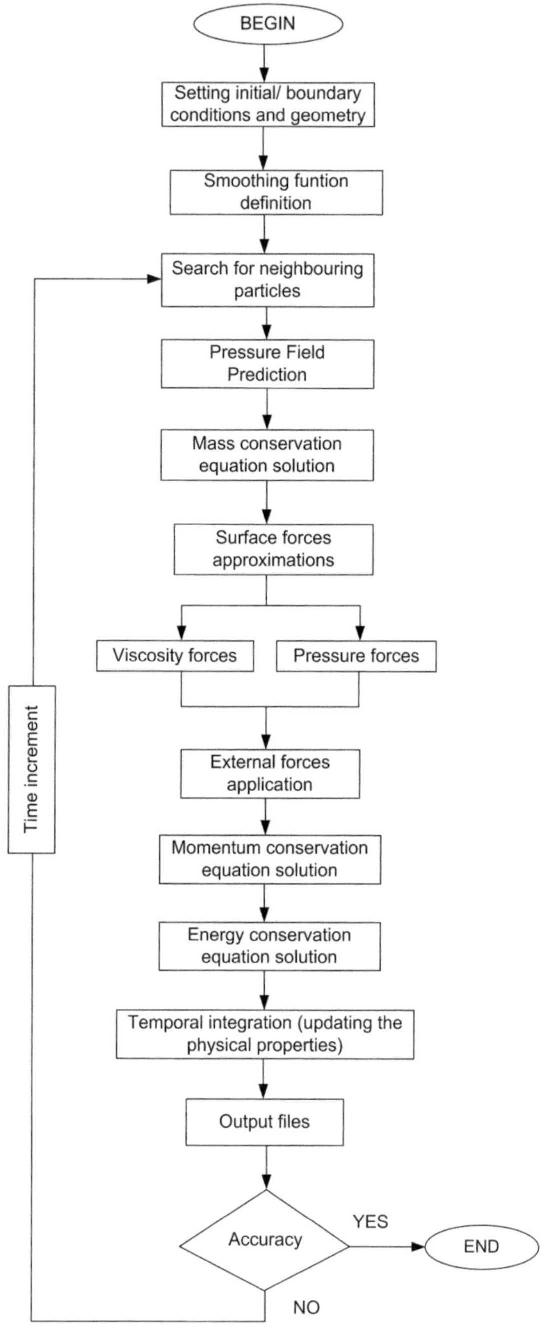

9. Energy conservation equation solution: the rate of change of specific energy in time is provided by the SPH method for all particles at domain.
10. Temporal integration: the updating of the properties of the particles (position, velocity, energy and others) is performed.
11. Output files: these files are obtained at the end of each numerical iteration, and from their data, graphical representations of the fluid's physical properties are generated.
12. Accuracy: this routine verifies whether the desirable accuracy has been achieved or if a new iteration will have to be executed. In the latter case, we must return to step 3 (search for the neighbouring particles) and follow the remaining steps until a new verification of the accuracy of numerical results is performed.

In steps 5 and 6, the correction of the physical properties of the particles (density and pressure gradient) located near the boundaries can be made using the Corrective Smoothed Particle Method (CSPM), as explained in Sect. 3.4.1.

The flowchart of the general SPH code is presented in the figure. A more detailed presentation of this code is found in Fraga Filho [1] (Fig. 1).

Conclusion

At the end of this work and in the form of a conclusion, the author would like to offer a few words to the reader.

This book aimed to present the SPH particle method fundamentals and basic applications in continuum mechanics. A presentation of the physical-mathematical modelling was carried out, as well as the SPH method, its consistency, numerical aspects and techniques of treatment of the boundaries and interfaces.

Throughout the presentation of the contents, critical observations on the use of molecular concepts in the continuum scale were performed. The comments, intended for mathematicians, physicists, chemists, engineers, students and researchers in general, are questionings and some conclusions, on the techniques currently employed in the continuum scale.

The limit between the molecular and continuous scales and, consequently, the validity of microscopic concepts and models employed in the treatment of macroscopic problems were discussed. Without the desire of simply criticizing solutions already presented in the literature, alternative solutions, faithful to the concepts of the continuum mechanics, have been proposed and discussed.

In recent years, a great scientific effort is being made aiming to bridge the gap between molecular and continuum approaches, in order to solve problems in minor scales. The understanding and modelling of many phenomena require the development of studies in a challenging interface area, where occurs the intersection of multiple disciplines, such as physics, chemistry, material sciences and biology [2].

Hybrid molecular continuum, multiscale [3,4] and dissipative particle dynamics (DPD) models [5,6] are recent computational methods that couple the events in the continuum and molecular/atomistic domains.

In the nanoscopic scale, the fluctuations in the averaged properties (due to the motion and behaviour of individual particles) begin to have a significant effect on the behaviour of a system and must be taken into account in the context of any particular problem. Scientific challenges appear when moving away from the macroscopic

C. A. Dutra Fraga Filho, *Smoothed Particle Hydrodynamics*,
https://doi.org/10.1007/978-3-030-00773-7

domain and approaching the molecular scale. The behaviour of nanoparticles is not only different from macroscopic particles, but they do not obey the same physical laws. A mesoscopic region separates the laws of continuum mechanics and quantum physics. A gap has been left in the full range of scaling from macro to nano. Reference [7] makes a basic presentation on the mesoscale and its characteristics.

The problems discussed in this study do not involve the two scales (continuum and molecular); on the contrary, they are defined in the macroscopic scale. In physical and engineering problems defined in the continuum scale, these models cannot be employed. It is always necessary to obey the classical continuum mechanics laws and not to mix them with the laws governing the molecular scale (quantum physics and molecular dynamics). The computational solutions obtained in the continuum domain using fictitious particles/artificial repulsive forces are not related to hybrid or multiscale methods. They are only results obtained from purely computational techniques in which the empirical calibration of parameters, such as those in Lennard-Jones molecular force (Eq. (3.113)), is performed.

In the process of implementing the numerical code used for the simulation of oil spreading in the calm sea (section X), we adopted the methodology to verify the results provided by the SPH approximation for each term of the physical conservation equations from the comparison with analytical results, experimental or existing in the literature, on the three problems studied in previous stages of the study of oil spreading. That is, the conclusions and discussions presented in this book are based on the theoretical foundations of classical physics and continuous mechanics, but also the computational implementation and verification of the attainable results in each stage of scientific research. The author feels obliged to present this clarification, out of respect for those who develop research on methods of particles applied in the domain of the continuum.

I sincerely hope that scientific research continues in its progress of proposing solutions to the problems that arise in the most diverse knowledge areas. But I also hope that answers to the questions and doubts that wander in the minds of scientists, students and researchers are sought, taking into account that the problems must be correctly modelled and the results achieved are consistent with the physical or chemical phenomena studied.

May this work be an incentive to the incessant search for answers, which is the essence of the scientific advancement.

Appendix A
Smoothing Functions, Derivatives and Normalization Constants

In this additional material, the interpolation functions used in numerical simulations, their derivatives and normalization constants, which satisfy the condition of unity (Sect. 3.1), will be presented.

- **Lucy's Quartic Kernel** (Fig. A.1)

  In the polar coordinate system:

$$W(q,h) = \alpha_D \begin{cases} (1+3q)\left(1-q^3\right), & 0 \le q \le 1 \\ 0, & \text{in the other case.} \end{cases}$$

where

r is the position of the fixed point
r$'$ is the position of the variable point

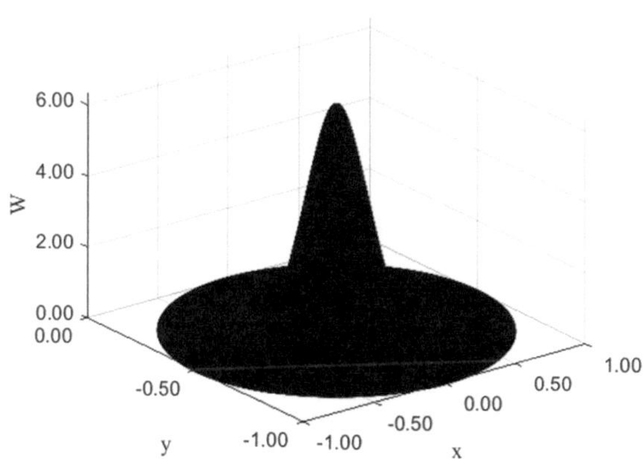

Fig. A.1 Lucy's quartic kernel in a two-dimensional domain

C. A. Dutra Fraga Filho, *Smoothed Particle Hydrodynamics*,
https://doi.org/10.1007/978-3-030-00773-7

$q = \frac{\left|(\mathbf{r} - \mathbf{r}', h)\right|}{h}$ and α_D is the kernel's normalization constant.

Derivative:

$$\frac{\partial W(q, h)}{\partial q} = \alpha_D \left(-12q + 24q^2 - 12q^3\right)$$

Normalization Constants:

(1) In a 2-D Domain:

$\mathbf{0.0 \le q \le 1.0}$:

$$\int_{\Omega} W d\Omega = 1 \text{ (Unity property)}$$

$$\alpha_D \int_0^{2\pi} \int_0^h (1 + 3q)(1 - q)^3 r dr d\theta = 1.$$

$$\therefore \boxed{\alpha_D = \frac{5}{\pi h^2}}$$

(2) In a 3-D Domain:

Assuming a spherical coordinate system and integrating in the first octant, we have:

$\mathbf{0.0 \le q \le 1.0}$:

$$\int_0^{\frac{\pi}{2}} \int_0^{\frac{\pi}{2}} \int_0^h (1 + 3q)\,(1 - q)^3 r^2 \sin\theta dr d\theta d\phi = \frac{2\pi h^3}{105}$$

In the whole domain of influence (all octants):

$$\int_{\Omega} W d\Omega = 1 \text{ (Unity property)}$$

$$\int_{\Omega} W d\Omega = \alpha_D \left[8 \left(\frac{2\pi h^3}{105}\right)\right] = 1$$

$$\therefore \boxed{\alpha_D = \frac{105}{16\pi h^3}}$$

- **Cubic Spline Kernel** (Fig. A.2)

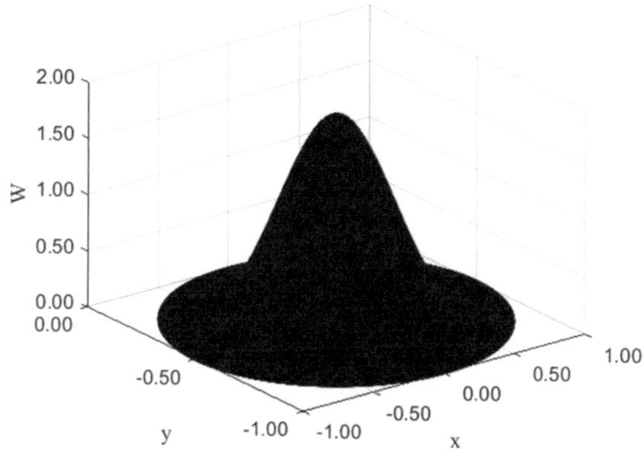

Fig. A.2 Cubic spline kernel in a two-dimensional domain

$$W(q,h) = \alpha_D \begin{cases} \left(\dfrac{2}{3} - q^2 + \dfrac{1}{2}q^3\right), & 0 \le q \le 1 \\ \left[\dfrac{1}{6}(2-q)^3\right], & 1 < q \le 2 \\ 0, & \text{in the other case.} \end{cases}$$

Derivatives:

$0.0 \le q \le 1.0$:

$$\frac{\partial W(q,h)}{\partial q} = \alpha_D\left(-2q + \frac{3q^2}{2}\right)$$

$1.0 < q \le 2.0$:

$$\frac{\partial W(q,h)}{\partial q} = \alpha_D\left(\frac{-(2-q)^2}{2}\right)$$

Normalization Constants:

(1) In a 2-D Domain:

$0.0 \le q \le 1.0$:

$$\int_0^{2\pi}\int_0^{h}\left(\frac{2}{3} - q^2 + \frac{1}{2}q^3\right) r dr d\theta = \frac{11\pi h^2}{30}$$

$1.0 < q \le 2.0$:

$$\int_0^{2\pi}\int_h^{2h}\frac{1}{6}(2-q)^3\, r dr d\theta = \frac{3\pi h^2}{30}$$

In the whole domain of influence:

$$\int_{\Omega} W d\Omega = 1 \text{ (Unity property)}$$

$$\int_{\Omega} W d\Omega = \alpha_D \left(\frac{11\pi h^2}{30} + \frac{3\pi h^2}{30} \right) = 1$$

$$\therefore \boxed{\alpha_D = \frac{15}{7\pi h^2}}$$

(2) In a 3-D Domain:

Integrating in the first octant:

$0.0 \leq q \leq 1.0$:

$$\int_0^{\frac{\pi}{2}} \int_0^{\frac{\pi}{2}} \int_0^{h} \left(\frac{2}{3} - q^2 + \frac{1}{2} q^3 \right) r^2 \sin\theta dr d\theta d\phi = \frac{19\pi h^3}{360}$$

$1.0 < q \leq 2.0$:

$$\int_0^{\frac{\pi}{2}} \int_0^{\frac{\pi}{2}} \int_h^{2h} \frac{1}{6} (2 - q)^3 r^2 \sin\theta dr d\theta d\phi = \frac{11\pi h^3}{360}$$

In the whole domain of influence (all octants):

$$\int_{\Omega} W d\Omega = 1 \text{ (Unity property)}$$

$$\int_{\Omega} W d\Omega = \alpha_D \left[8 \left(\frac{19\pi h^3}{360} + \frac{11\pi h^3}{360} \right) \right] = 1$$

$$\therefore \boxed{\alpha_D = \frac{3}{2\pi h^3}}$$

- **New Quartic Kernel** (Fig. A.3)

$$W(\mathbf{r} - \mathbf{r}', h) = \alpha_D \begin{cases} \left(\frac{2}{3} - \frac{9}{8} q^2 + \frac{19}{24} q^3 - \frac{5}{32} q^4 \right), & 0 \leq q \leq 2 \\ 0, & \text{in the other case.} \end{cases}$$

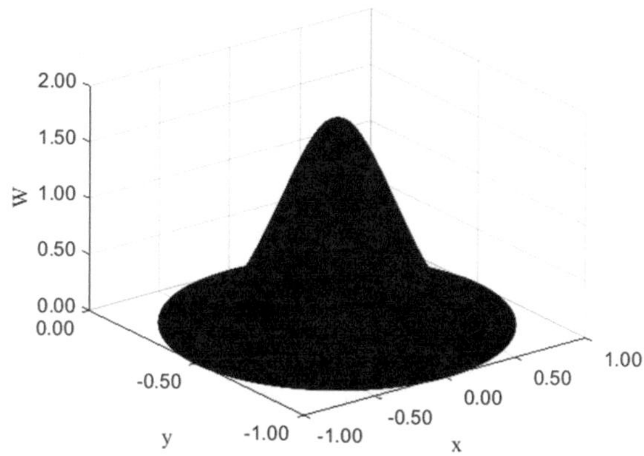

Fig. A.3 New quartic kernel in a two-dimensional domain

Derivative:

$0.0 \leq q \leq 2.0$:

$$\frac{\partial W(q,h)}{\partial q} = \alpha_D\left(-\frac{18q}{8} + \frac{57q^2}{24} - \frac{20q^3}{32}\right)$$

Normalization Constants:

(1) In a 2-D Domain:

$0.0 \leq q \leq 2.0$:

$$\int_0^{2\pi}\int_h^{2h}\left(\frac{2}{3} - \frac{9}{8}q^2 + \frac{19}{24}q^3 - \frac{5}{32}q^4\right) r\,dr\,d\theta = \frac{7\pi h^2}{15}$$

In the whole domain of influence:

$$\int_\Omega W d\Omega = 1 \text{ (Unity property)}$$

$$\int_\Omega W d\Omega = \alpha_D\left(\frac{11\pi h^2}{30} + \frac{3\pi h^2}{30}\right) = 1$$

$$\therefore \boxed{\alpha_D = \frac{15}{7\pi h^2}}$$

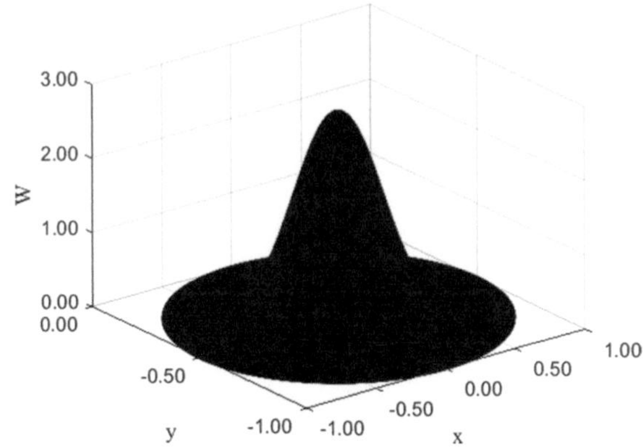

Fig. A.4 Quintic spline kernel in a two-dimensional domain

(2) In a 3-D Domain:

Integrating in the first octant:

$\mathbf{0.0 \le q \le 2.0}$:

$$\int_0^{\frac{\pi}{2}}\int_0^{\frac{\pi}{2}}\int_0^{2h}\left(\frac{2}{3}-\frac{9}{8}q^2+\frac{19}{24}q^3-\frac{5}{32}q^4\right)r^2\sin\theta drd\theta d\phi=\frac{52}{630\pi h^3}$$

In the whole domain of influence (all octants):

$$\int_\Omega Wd\Omega = 1 \text{ (Unity property)}$$

$$\int_\Omega Wd\Omega = \alpha_D\left[8\left(\frac{416\pi h^2}{630}\right)\right] = 1$$

$$\therefore \boxed{\alpha_D = \frac{315}{208\pi h^3}}$$

- **Quintic Spline Kernel** (Fig. A.4)

$$W(\mathbf{r}-\mathbf{r}',h)=\alpha_D\begin{cases}(3-q)^5-6(2-q)^5+15(1-q)^5, & 0\le q\le 1\\ (3-q)^5-6(2-q)^5, & 1<q\le 2\\ (3-q)^5, & 2<q\le 3\\ 0, & \text{in the other case.}\end{cases} \tag{A.1}$$

Derivatives:

$\mathbf{0.0 \leq q \leq 1.0}$:

$$\frac{\partial W(q,h)}{\partial q} = \alpha_D \left(-120q + 120q^3 - 50q^4\right)$$

$\mathbf{1.0 < q \leq 2.0}$:

$$\frac{\partial W(q,h)}{\partial q} = \alpha_D \left(75 - 420q + 450q^2 - 180q^3 + 25q^4\right)$$

$\mathbf{2.0 < q \leq 3.0}$:

$$\frac{\partial W(q,h)}{\partial q} = \alpha_D \left(405 + 540q - 270q^2 + 60q^3 - 5q^4\right)$$

Normalization Constants:

(1) In a 2-D Domain:

$\mathbf{0.0 \leq q \leq 1.0}$:

$$\int_0^{2\pi}\int_0^{h} \left[(3-q)^5 - 6(2-q)^5 + 15(1-q)^5\right] r dr d\theta = \frac{302\pi h^2}{7}$$

$\mathbf{1.0 \leq q \leq 2.0}$:

$$\int_0^{2\pi}\int_h^{2h} \left[(3-q)^5 - 6(2-q)^5\right] r dr d\theta = \frac{171\pi h^2}{7}$$

$\mathbf{2.0 \leq q \leq 3.0}$:

$$\int_0^{2\pi}\int_{2h}^{3h} (3-q)^5 r dr d\theta = \frac{5\pi h^2}{7}$$

In the whole domain of influence:

$$\int_\Omega W d\Omega = 1 \text{ (Unity property)}$$

$$\int_\Omega W d\Omega = \alpha_D \left(\frac{302\pi h^2}{7} + \frac{171\pi h^2}{7} + \frac{5\pi h^2}{7}\right) = 1$$

$$\therefore \boxed{\alpha_D = \frac{7}{478\pi h^2}}$$

(2) In a 3-D Domain:

Integrating in the first octant:

$0.0 \le q \le 1.0$:

$$\int_0^{\frac{\pi}{2}}\int_0^{\frac{\pi}{2}}\int_0^{h}\left[(3-q)^5 - 6(2-q)^5 + 15(1-q)^5\right]r^2 \sin\theta dr d\phi = \frac{730\pi h^3}{14}$$

$1.0 \le q \le 2.0$:

$$\int_0^{\frac{\pi}{2}}\int_0^{\frac{\pi}{2}}\int_h^{2h}\left[(3-q)^5 - 6(2-q)^5\right]r^2 \sin\theta dr d\phi = \frac{907\pi h^3}{14}$$

$2.0 \le q \le 3.0$:

$$\int_0^{\frac{\pi}{2}}\int_0^{\frac{\pi}{2}}\int_{2h}^{3h}(3-q)^5 r^2 \sin\theta dr d\phi = \frac{43\pi h^3}{14}$$

In the whole domain of influence (all octants):

$$\int_\Omega W d\Omega = 1 \text{ (Unity property)}$$

$$\int_\Omega W d\Omega = \alpha_D\left[8\left(\frac{730\pi h^3}{14} + \frac{907\pi h^3}{14} + \frac{43\pi h^3}{14}\right)\right] = 1$$

$$\therefore \boxed{\alpha_D = \frac{1}{120\pi h^3}}$$

Comments

The integration has been performed over the domain of influence with a unity radius (kh = 1.0, where k is a scaling factor that depends on the smoothing function).

The scaling factor divides the domain of influence in regions, in which the smoothing function presents different mathematical expressions, and defines the number of regions on which the integration is performed in order to obtain the normalization constant.

The unity property does not depend on the length of the support radius (kh).

The kernel's unit, not shown in the deductions, is the inverse unit of volume.

Appendix B
Deduction of the SPH Laplacian Operator

In this appendix, the mathematical deduction of the Laplacian of a function (in a 2-D domain) using the SPH method will be presented.

The expression obtained has been used to approximate the terms of viscous forces in the momentum conservation equation and the Laplacian of the temperature (at the study of the heat diffusion in a flat plate).

For the approximation of the Laplacian of a function, the expansion of the Taylor series was used. In a two-dimensional domain, it is possible to determine the value of a function at a point $X' = (x', y')$, around a fixed point $X = (x, y)$:

$$f\left(x', y'\right) = f(x, y) + \left(x' - x\right)\frac{\partial f}{\partial x}\bigg|_{(x,y)} +$$

$$\left(y' - y\right)\frac{\partial f}{\partial y}\bigg|_{(x,y)} + \frac{1}{2}\left[\left(x' - x\right)^2\frac{\partial^2 f}{\partial x^2}\bigg|_{(x,y)} +\right.$$

$$\left.\left(y' - y\right)^2\frac{\partial^2 f}{\partial y^2}\bigg|_{(x,y)}\right] + \left(x' \quad x\right)\left(y' - y\right)\frac{\partial^2 f}{\partial x \partial y}\bigg|_{(x,y)} +$$

$$R_n\left(X' - X\right) \tag{B.1}$$

where $R_n\left(X' - X\right)$ is the remainder of the Taylor series.

Assuming an error of order 3 in the approximation, multiplying Eq. (B.1), truncated, by $\frac{\left(X - X'\right)}{\left|X - X'\right|^2}\nabla W(X - X', h)$, and integrating it:

$$\int_{\Omega} \overbrace{f\left(x', y'\right)\Delta X \cdot \nabla W(X - X', h) dX'}^{A} =$$

C. A. Dutra Fraga Filho, *Smoothed Particle Hydrodynamics*,
https://doi.org/10.1007/978-3-030-00773-7

$$\int_{\Omega} \overbrace{f\,(\mathrm{x},\mathrm{y})\,\Delta X\cdot\nabla \mathrm{W}(\mathrm{X}-\mathrm{X}^{'},h)\mathrm{dX}'}^{B}+\int_{\Omega}\overbrace{\left(\mathrm{x}^{'}-\mathrm{x}\right)\left.\frac{\partial f}{\partial \mathrm{x}}\right|_{(\mathrm{x},\mathrm{y})}\Delta X\cdot\nabla \mathrm{W}(\mathrm{X}-\mathrm{X}^{'},h)\mathrm{dX}'}^{C}+$$

$$\int_{\Omega}\overbrace{\left(\mathrm{y}^{'}-\mathrm{y}\right)\left.\frac{\partial f}{\partial \mathrm{y}}\right|_{(\mathrm{x},\mathrm{y})}\Delta X\cdot\nabla \mathrm{W}(\mathrm{X}-\mathrm{X}^{'},h)\mathrm{dX}'}^{D}+$$

$$\int_{\Omega}\overbrace{\frac{1}{2}\left(\left(\mathrm{x}^{'}-\mathrm{x}\right)^{2}\left.\frac{\partial^{2} f}{\partial \mathrm{x}^{2}}\right|_{(\mathrm{x},\mathrm{y})}+\left(\mathrm{y}^{'}-\mathrm{y}\right)^{2}\left.\frac{\partial^{2} f}{\partial \mathrm{y}^{2}}\right|_{(\mathrm{x},\mathrm{y})}\right)\Delta X\cdot\nabla \mathrm{W}(\mathrm{X}-\mathrm{X}^{'},h)\mathrm{dX}'}^{E}+$$

$$\int_{\Omega}\overbrace{\frac{1}{2}\left(\mathrm{x}^{'}-\mathrm{x}\right)\left(\mathrm{y}^{'}-\mathrm{y}\right)\left.\frac{\partial^{2} f}{\partial \mathrm{x}\partial \mathrm{y}}\right|_{(\mathrm{x},\mathrm{y})}\Delta X\cdot\nabla \mathrm{W}(\mathrm{X}-\mathrm{X}^{'},h)\mathrm{dX}'}^{F} \tag{B.2}$$

where $\Delta X = \dfrac{\left(X - X^{'}\right)}{|X - X^{'}|^{2}}$.

The gradient of the kernel has the properties of anti-symmetry and non-normalization in the Euclidean coordinate system:

$$\boxed{\int_{\Omega}\left(X^{'}-X\right)_{p}\Delta X\cdot\nabla W\left(X^{'}-X,h\right)dX^{'}=0} \tag{B.3}$$

$$\boxed{\int_{\Omega}\left(X^{'}-X\right)_{l}\left(X^{'}-X\right)_{t}\Delta X\cdot\nabla W\left(X^{'}-X,h\right)dX^{'}=-\delta_{l,t}} \tag{B.4}$$

where $l = (1, 2)$, $t = (1, 2)$, $p = (1, 2)$, $\left(X^{'} - X\right)_1 = \left(\mathrm{x}^{'} - \mathrm{x}, 0\right)$, $\left(X^{'} - X\right)_2 = \left(0, \mathrm{y}^{'} - \mathrm{y}\right)$ and $\delta_{l,t}$ is the Kronecker delta function.

Applying the properties of the kernel, we conclude:

$$\int_{\Omega}\overbrace{\left(\mathrm{x}^{'}-\mathrm{x}\right)\left.\frac{\partial f}{\partial \mathrm{x}}\right|_{(\mathrm{x},\mathrm{y})}\Delta X\cdot\nabla \mathrm{W}(\mathrm{X}-\mathrm{X}^{'},h)\mathrm{dX}'=0}^{C}, \tag{B.5}$$

$$\int_{\Omega}\overbrace{\left(\mathrm{y}^{'}-\mathrm{y}\right)\left.\frac{\partial f}{\partial \mathrm{y}}\right|_{(\mathrm{x},\mathrm{y})}\Delta X\cdot\nabla \mathrm{W}(\mathrm{X}-\mathrm{X}^{'},h)\mathrm{dX}'=0}^{D}, \tag{B.6}$$

$$\int_{\Omega} \overbrace{\frac{1}{2}\left(\left(x^{'}-x\right)^{2}\frac{\partial^{2}f}{\partial x^{2}}\bigg|_{(x,y)}+\left(y^{'}-y\right)^{2}\frac{\partial^{2}f}{\partial y^{2}}\bigg|_{(x,y)}\right)\Delta X\cdot\nabla W(X-X^{'},h)dX'}^{E} = -\frac{1}{2}\left(\frac{\partial^{2}f}{\partial x^{2}}+\frac{\partial^{2}f}{\partial y^{2}}\right)\bigg|_{(x,y)}, \tag{B.7}$$

$$\int_{\Omega} \overbrace{\left(x^{'}-x\right)\left(y^{'}-y\right)\frac{\partial^{2}f}{\partial x\partial y}\bigg|_{(x,y)}\Delta X\cdot\nabla W(X-X^{'},h)dX'}^{F} = 0, \tag{B.8}$$

$$\boxed{\nabla^{2}f\big|_{(x,y)} = \left(\frac{\partial^{2}f}{\partial x^{2}}+\frac{\partial^{2}f}{\partial y^{2}}\right)\bigg|_{(x,y)} = -2\int_{\Omega} f[\left(x^{'},y^{'}\right)-f(x,y)]\Delta X\cdot\nabla W\left(X-X^{'},h\right)dX^{'}} \tag{B.9}$$

where $\nabla^{2}f\big|_{(x,y)}$ is the Laplacian of the function f, evaluated at the position X.

After the domain discretization by particles, we arrive at the following expression for the SPH approximation of the Laplacian of the function f in the position occupied by the fixed particle i:

$$\boxed{\nabla^{2}f_{i} = \left(\frac{\partial^{2}f_{i}}{\partial x^{2}}+\frac{\partial^{2}f_{i}}{\partial y^{2}}\right) = 2\sum_{j=1}^{n}\frac{m_{j}}{\rho_{j}}\left(f_{i}-f_{j}\right)\Delta X_{ij}\cdot\nabla W\left(X_{i}-X_{j},h\right)} \tag{B.10}$$

where

i is the fixed particle, where the Laplacian of the function f is being evaluated

j is each of the neighbouring particles of the fixed particle

$\nabla^{2}f_{i}$ is the SPH approximation of the Laplacian of the function f evaluated at the position occupied by the fixed particle $\Delta X_{ij} = \dfrac{X_{i}-X_{j}}{\left|X_{i}-X_{j}\right|^{2}}$.

In the polar coordinate system, the expression of the SPH Laplacian becomes:

$$\boxed{\nabla^{2}f_{i} = 2\sum_{j=1}^{n}\frac{m_{j}}{\rho_{j}}\left(f_{i}-f_{j}\right)\frac{\partial W(\mathbf{r}_{i}-\mathbf{r}_{j},h)}{\partial r}\frac{1}{\left|\mathbf{r}_{ij}\right|}} \tag{B.11}$$

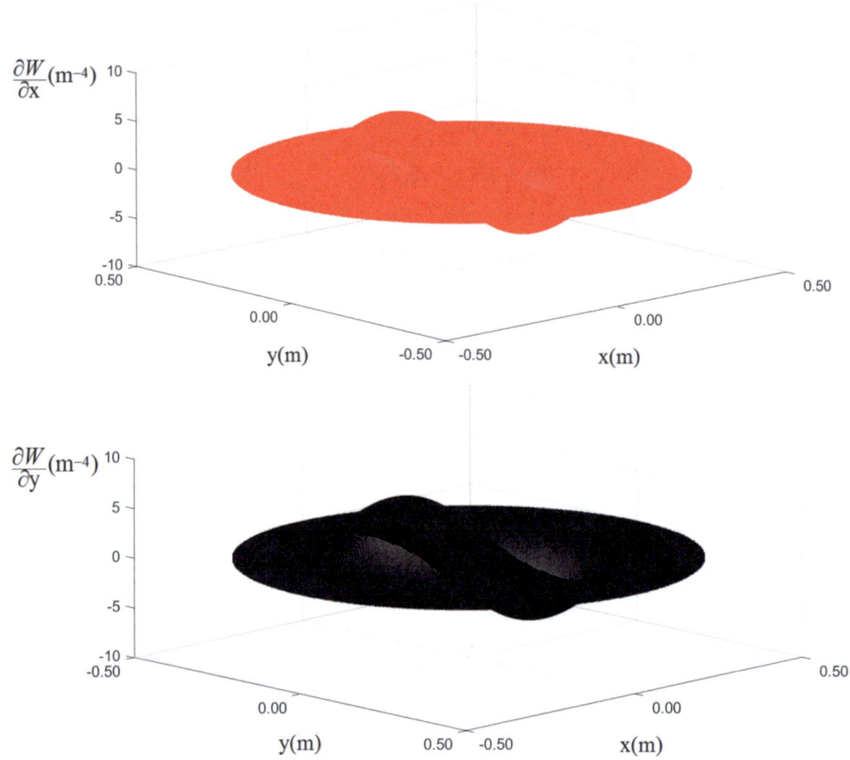

Fig. B.1 First derivatives of the kernel

where

r is the radial direction

$W(\mathbf{r}_i - \mathbf{r}_j, h)$ is the smoothing function written in polar coordinates evaluated at the position $(\mathbf{r}_i - \mathbf{r}_j)$

$\left|\mathbf{r}_{ij}\right| = \left|(\mathbf{r}_i - \mathbf{r}_j)\right|$ is the radial distance between the fixed and a neighbouring particle.

Figures B.1 and B.2 present the graphs of the kernel's derivatives in the following order: first derivatives (allowing the visualization of the anti-symmetry property) and second radial derivative. The kernel used was the cubic spline, but the behaviour of the graphs is the same for most of the smoothing functions used in the SPH method.

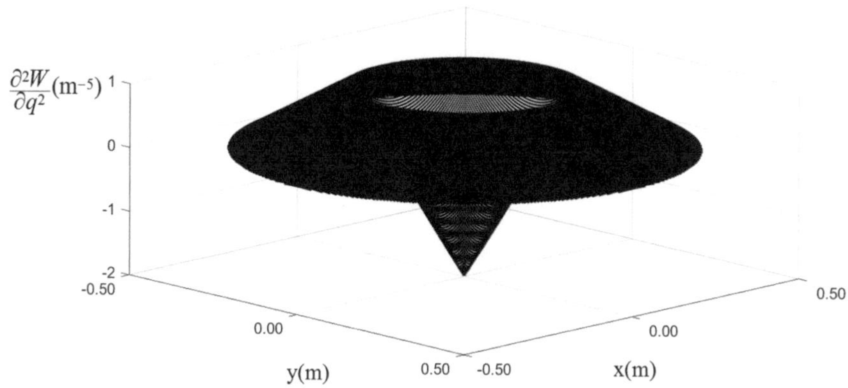

Fig. B.2 Second radial derivative of the kernel

FORTRAN Source Files

Heat Diffusion in a Homogeneous Flat Plate

The following are the open-source code (in FORTRAN Language) and a script in the MATLAB language for plotting the temperature images. Instructions for creating the executable program are provided at the beginning of the document.

```
PROGRAM flat_plate_diffusion

!-----------------------------------------------------------------
! These FORTRAN source files have been used in heat diffusion in a
! homogeneous flat plate (Sect. 4.2).
!-----------------------------------------------------------------
! The reader should follow the instructions below in order to
! create the program executable:
! 1. Within the working directory, create and open a new file in
! the FORTRAN Editor~
! 2. Copy and paste this PROGRAM and save as ''name.f90''
! 3. Create a folder called output within the working directory
! 4. Create a subfolder called temperature within the folder output
! 5. Compile and Run the program
! Output: in the file called TEMPERATURE_DIFFERENCES.DAT are the
! points where the smallest and the largest temperature differences
! occur (comparing the SPH results and solution provided by series).
!
!-----------------------------------------------------------------
! VARIABLES:
! aux_p - variable used in memory allocation
! dist - distance between the centres of mass of a fixed and a
! neighbour particle
! dt - time step (in seconds)
! dx - horizontal distance between the centres of mass of
! particles
! dy - vertical distance between the centres of mass of
! particles
! eps - accuracy of simulation
! finish - variable used to record the simulation time
```


C. A. Dutra Fraga Filho, *Smoothed Particle Hydrodynamics*,
https://doi.org/10.1007/978-3-030-00773-7

```
! hsml - smoothing lengths of particles
! int_function - interpolation function (kernel)
! itimestep - current iteration
! LX - length of the plate
! LY - width of the plate
! mass - mass of particles
! n_bound - number of boundary particles
! n_cells - number of cells in each Cartesian direction
! neighbour - matrix containing the neighbours of each particle
! of the domain
! n_global - sum of the number of particles at domain and
! boundary particles
! np_side - number of particles per side of the plate
! ntotal - total number of particles at domain
! response - variable used to verify the achievement of accuracy
! rho - densities of particles
! start - variable used to record the simulation time
! step_out - step for recording results
! support_radius - radius of the domain of influence
! time - simulation time
! T_0 - initial temperatures of particles
! T - temperature provided by SPH method at each numerical
! iteration
! TE - prescribed temperature at right side of the plate
! TN - prescribed temperature at top of the plate
! TS - prescribed temperature at bottom of the plate
! TW - prescribed temperature at left side of the plate
! temp_Laplacian - SPH Laplacian of temperature
! x - coordinates of particles
!-----------------------------------------------------------------
IMPLICIT NONE

INTEGER, DIMENSION (:,:), allocatable :: neighbour

DOUBLE PRECISION, DIMENSION (:), allocatable :: rho, mass, hsml, T_0, T,
temp_Laplacian
DOUBLE PRECISION, DIMENSION (:,:), allocatable :: x, dist, dx, dy
INTEGER int_function, aux_p, ntotal, itimestep, d, m, i, hours, minutes,
hours_seg, np_side, n_global, n_cells, n_bound, step_out
DOUBLE PRECISION TS, TN, TW, TE, LX, LY, xl, yl, factor
DOUBLE PRECISION start, finish, time, seconds, dt, eps, support_radius
CHARACTER(LEN=15) response

CALL cpu_time(start)

OPEN(01,file="output/domain_particles_positions.dat")
OPEN(02,file="output/boundary_particles_positions.dat")
OPEN(03,file="output/initial_temperature.dat")

!------------------------Plate geometry---------------------------
LX=1.0
LY=1.0
!-----------------------------------------------------------------

!-----------------------------------------------------------------
!---------------DEFINING THE TEMPERATURES AT BOUNDARIES-----------
```

```
!------------------(DIRICHLET BOUNDARY CONDITIONS)----------------
!-----------------------------------------------------------------

TS=100.0 !Temperature at the lower boundary
TN=0.0 !Temperature at the upper side boundary
TW=0.0 !Temperature at the left side boundary
TE=0.0 !Temperature at the right side boundary

!-------------------Time step and Accuracy-----------------------
dt = 1.e-5
eps = 1.e-6
!-----------------------------------------------------------------

print*, ' '
print*, 'Enter with the number of particles per side of the plate:'
print*, ' '
read*, np_side
print*, ' '
print*, 'Choose the interpolation Function (Kernel):'
print*, ' '
print*, '1 - Lucy''s Quartic Kernel'
print*, '2 - Cubic Spline Kernel'
print*, '3 - New Quartic Kernel'
print*, '4 - Quintic Spline Kernel'
print*, ' '
read*, int_function
print*, ' '
print*, 'Enter with the step for the output files:'
print*, ' '
read*, step_out

!Distance between the centres of mass

aux_p = np_side**2 + 1500 !total number of particles (for memory
allocation)

allocate (neighbour(aux_p,150), rho(aux_p), hsml(aux_p),
T_0(aux_p), T(aux_p), temp_Laplacian(aux_p), x(aux_p,2),
dist(aux_p,150), dx(aux_p,150), dy(aux_p,150), mass(aux_p))

response='non_convergence'
itimestep=1

print*, ' '
print*,'-------------------------------------------------------'
print*,'--FLAT PLANE DIFFUSION SIMULATION USING THE SPH METHOD-'
print*,'-------------------------------------------------------'
print*, ' '

dx = LX/np_side
dy = LY/np_side

CALL temperature_series(aux_p, np_side, TS, TN, TW, TE, dx, dy, dt)

CALL input(aux_p, x, mass, rho, np_side, ntotal, T_0, LX, LY,
support_radius, n_cells)
```

```
CALL boundary_particles(aux_p, itimestep, np_side, TS, TN, TW, TE, LX, LY,
dx, dy, n_bound, ntotal, n_global, mass,x,rho,T_0)

CALL neighbour_search(aux_p, x, n_global, ntotal, n_bound, neighbour, dx,
dy, support_radius, dist)

!-----------------------------------------------------------------
!--------------------UNTIL THE ACCURACY IS ACHIEVED---------------
!-----------------------------------------------------------------

DO WHILE (response.eq.'non_convergence')

  CALL SPH_Temperature_Laplacian (aux_p, int_function, factor, hsml,
ntotal, dist, mass, rho, support_radius, neighbour, T_0, temp_Laplacian,
itimestep)

  CALL temporal_integration(aux_p, itimestep, x, ntotal, dt,
temp_Laplacian, T_0, T, step_out)

  CALL convergence (aux_p, itimestep,T_0,T,eps,response, ntotal)

  CALL temperature_differences(itimestep, dt, ntotal, step_out)

  IF (response.eq.'convergence') THEN
    PRINT*, ' ACCURACY ACHIEVED IN THE ', itimestep, 'a. ITERATION'
    PRINT*, ' END OF SIMULATION '
  END IF

END DO

CALL cpu_time(finish)

time=finish-start
hours_seg=3600; !hours em seconds
hours = (time/hours_seg) !resultado da hora
minutes = (time -(hours_seg*hours))/60
seconds = (time -(hours_seg*hours)-(minutes*60))
print*,' '
print*, ' ----------------------------------------------------'
WRITE (*,*) 'TIME PROCESSING (CPU TIME) '
WRITE (*,*) hours ,' hours', minutes,' minutes', seconds, ' seconds.'
print*, ' ----------------------------------------------------'

deallocate(neighbour, rho, hsml, T_0, T, temp_Laplacian, x, dist, dx, dy, mass)

OPEN(776,file= 'output/SIMULATION_PARAMETERS.DAT')
REWIND(776)
WRITE (776,*) ' '
IF (int_function.EQ.1) THEN !--------LUCY'S QUARTIC KERNEL----
  WRITE (776,*) 'Interpolation Function: Lucy''s Quartic Kernel'
  WRITE(776,*) ' '
ELSE
  IF (int_function.EQ.2) THEN !------CUBIC SPLINE KERNEL-------
   WRITE (776,*) 'Interpolation Function: Cubic Spline Kernel'
    WRITE(776,*) ' '
  ELSE
    IF (int_function.EQ.3) THEN !----NEW QUARTIC KERNEL--------
```

```
      WRITE (776,*) 'Interpolation Function: New Quartic Kernel'
      WRITE(776,*) ' '
    ELSE
      IF (int_function.EQ.4) THEN !-----QUINTIC SPLINE KERNEL---
        WRITE (776,*) 'Interpolation Function: Quintic Spline Kernel'
        WRITE(776,*) ' '
      END IF
    END IF
  END IF
END IF

OPEN(25,FILE='NUMBER_OF_ITERATIONS.DAT')
WRITE(25,*) itimestep
 CLOSE(25)

WRITE (776,*) 'Number of particles per side of plate = ', np_side
WRITE (776,*) ' '
WRITE (776,*) 'Time step (dt, em seconds) = ', dt
WRITE (776,*) ' '
WRITE (776,*) 'Accuracy ( abs(T(m+1) - T(m)) ) = ', eps
WRITE (776,*) ' '
WRITE (776,*) 'Time for the physical diffusion =', itimestep*dt
WRITE (776,*) ' '
WRITE (776,*) 'Time processing (CPU time) = ', hours ,' hours',
minutes,' minutes', seconds, ' seconds.'

 CLOSE(01)
 CLOSE(02)
 CLOSE(03)
 CLOSE(776)

END PROGRAM

!******************** SOLUTION BY SERIES ************************

SUBROUTINE temperature_series(aux_p, np_side, TS, TN, TW, TE, dx, dy, dt)

!-----------------------------------------------------------------
! This subroutine defines the positions of the centres of mass of
! particles at domain and makes the calculation of thc
! temperatures at steady-state using the solution by series
!
! x - coordinates of particles
! mass - mass of particles
! KMAX - number of terms in series
!-----------------------------------------------------------------

IMPLICIT NONE

DOUBLE PRECISION, DIMENSION (:,:), ALLOCATABLE :: T , X1
DOUBLE PRECISION, DIMENSION (:), ALLOCATABLE :: POSX, POSY, TEMP
DOUBLE PRECISION LX, LY, dx, dy, aux, TS, TN, TW, TE, X, Y, SINX, Txy, PI,
dt
INTEGER aux_p, np_side, I, d, J, K, MP1, NP1, KMAX, cont, n_bound

MP1=np_side + 1
NP1=np_side + 1
```

```
ALLOCATE( T(MP1,NP1), POSX(aux_p), POSY(aux_p), TEMP(aux_p), X1(1500,3))

PI=ACOS(-1.)
KMAX=90

OPEN(UNIT=10,FILE='output/TEMPERATURES.DAT')
OPEN(UNIT=11,FILE='output/VERIFYING_TEMP_SERIES.DAT')

!-------------------------------------------------------------
!Definition of positions occupied by centres of mass of particles

cont=0
DO I = 1,np_side
  X =(I-1)*dx + dx/2

  DO J = 1,np_side
    cont = cont + 1
    Y = (J-1)*dy + dy/2
    POSX(cont)= X
    POSY(cont)= Y

    !Calculation of temperatures

     Txy=0.
     DO K = 1, KMAX, 2
       SINX = SIN(K*PI*X)
       aux = ((SINH(K*PI*(Y-1.))) /(SINH(K*PI)))
       Txy = Txy + ((-4.*TS)/(K*PI))* SINX * aux
     END DO

     TEMP(cont)=Txy

   END DO

 END DO

 DO I = 1,cont
   WRITE(11,23) I, POSX(I), POSY(I), TEMP(I)
 END DO
23 FORMAT (1X,I10, 1X, D21.14, 1X, D21.14, 1X, D21.14)
 close(11)
 OPEN(23,FILE='N_PART_SERIES.DAT')
 WRITE(23,*) cont
 WRITE(23,*) dt
 CLOSE(23)

!Defining a line of boundary particles and temperatures
 n_bound = 0
 J = 1
 DO I = 1, MP1
   n_bound= n_bound+1
   X1(n_bound,1)=(I-1)*dx
   X1(n_bound,2)=0.
   X1(n_bound,3)=TS !Lower boundary
 END DO
```

```
J = NP1
DO I = 1,MP1
  n_bound = n_bound+1
  X1(n_bound,1)=(I-1)*dx
  X1(n_bound,2)=(NP1-1)*dy
  X1(n_bound,3)=TN !Upper boundary
END DO

I=1
DO J=2,np_side
n_bound = n_bound+1
   X1(n_bound,1)=0.
   X1(n_bound,2)=(J-1)*dy
   X1(n_bound,3)=TW !Left boundary
 END DO

 I=MP1
 DO J=2,np_side
  n_bound=n_bound+1
  X1(n_bound,1)=(MP1-1)*dx
  X1(n_bound,2)=(J-1)*dy
  X1(n_bound,3)=TE !Right boundary
END DO

OPEN(12,FILE='output/BOUNDARY_PARTICLES.DAT')
DO I=1,n_bound
  WRITE(12,23) I, (X1(I,d),d=1,3)
END DO
 CLOSE(12)

END SUBROUTINE temperature_series

!*************************** INPUT *******************************

SUBROUTINE input(aux_p, x, mass, rho, np_side, ntotal, T_0, LX, LY,
support_radi us, n_cells)

!-----------------------------------------------------------------
! This subroutine makes the distribution of particles at domain
! and the initial properties are set
!
! x - positions of particles
! mass - mass of particles
! rho - densities of particles
! hsml - smoothing lengths of particles
! ntotal - total number of particles
! ncells - total number of cells in each direction
! T_0 - initial temperature

IMPLICIT NONE
INTEGER np_side, aux_p , n_cells, ntotal
DOUBLE PRECISION LX, LY, xl, yl, dx, dy, support_radius
INTEGER i, k, d
DOUBLE PRECISION mass(aux_p), rho(aux_p), T_0(aux_p)
DOUBLE PRECISION x(aux_p,2), M(aux_p,4)
```

```
n_cells=np_side/5
yl= LY
xl = LX
dx = xl/np_side
dy = yl/np_side
OPEN(23,FILE='N_PART_SERIES.DAT')
READ (23,*) k
 CLOSE(23)

OPEN(UNIT=11,FILE='output/VERIFYING_TEMP_SERIES.DAT') !PARTICLES AT DOMAIN
DO i=1,k
  READ(11,*) (M(i,d),d=1,4)
END DO

DO i=1,k
  x(i,1)=M(i,2)
  x(i,2)=M(i,3)
END DO

 CLOSE (11)

!--------------------DEFINING THE SUPPORT RADIUS ---------------
support_radius = (xl/real(n_cells))/2.
!-----------------------------------------------------------

!--------------------DEFINING THE PROPERTIES OF PARTICLES -------

DO i = 1, aux_p
  rho (i) = 1.
  mass(i) = dx*dy*rho(i)
  T_0(i) = 0.
  WRITE (03,654) T_0(i)
END DO

654 format(1x,D21.14)

DO i=1, k
  WRITE (01,1013) (x(i,d), d = 1, 2)
END DO

1013 format(2(1x,D21.14))
ntotal = k

WRITE (*,*)'-----------------------------------------------'
WRITE (*,*)' TOTAL NUMBER OF PARTICLES AT DOMAIN : '
WRITE (*,*) ntotal
WRITE (*,*)'-----------------------------------------------'
WRITE (*,*) ' '
WRITE (*,*) ' PRESS ANY KEY TO CONTINUE... '

pause

END SUBROUTINE input
```

```
!************************ BOUNDARY PARTICLES*********************

SUBROUTINE boundary_particles(aux_p, itimestep, np_side, TS, TN, TW, TE,
LX, LY, dx, dy, n_bound, ntotal, n_global, mass,x,rho,T_0)
!-----------------------------------------------------------------
! This subroutine defines a line of particles at the contour of
! the flat plate and the Dirichlet boundary conditions
!
! itimestep - current time step
! ntotal - total number of particles at domain
! n_bound - number of boundary particles
! n_global - sum of the number of particles at domain and
! boundary particles
! hsml - smoothing length
! mass - masses of partic,ces
! x - positions of particles
! rho - densities of particles
! T_0 - initial temperature
!-----------------------------------------------------------------

IMPLICIT NONE

INTEGER aux_p, n_global, np_side, ntotal, n_bound, itimestep
INTEGER i, j, d, k
DOUBLE PRECISION TS, TN, TW, TE, LX, LY, dx,dy,MAX_X, MAX_Y
DOUBLE PRECISION :: rho(aux_p), mass(aux_p), hsml(aux_p), T_0(aux_p)
DOUBLE PRECISION :: x(aux_p,2)

n_bound = 0

!-----------Defining a line of particles on the Upper side and setting the
physical properties

DO i = 1, 2*np_side -1
  n_bound = n_bound + 1
  x(ntotal + n_bound,1) = i*dx/2
  x(ntotal + n_bound,2) = LY
  rho (ntotal + n_bound) = 1.
  mass(ntotal + n_bound) = rho (ntotal + n_bound) * dx * dy
  T_0(ntotal + n_bound)= TN
END DO

! -----------Defining a line of particles on the Lower side and setting the
physical properties
 DO i = 1, 2*np_side -1
   n_bound = n_bound + 1
   x(ntotal + n_bound,1) = i*dx/2
   x(ntotal + n_bound,2) = 0.
   rho (ntotal + n_bound) = 1.
   mass(ntotal + n_bound) = rho (ntotal + n_bound) * dx * dy
   T_0(ntotal + n_bound)= TS
 END DO

!----------------Defining a line of particles on the Left side and setting
the physical properties
 DO i = 1, 2*np_side + 1
   n_bound = n_bound + 1
```

```
   x(ntotal + n_bound,1) = 0.
   x(ntotal + n_bound,2) = (i-1)*dx/2
   rho (ntotal + n_bound) = 1.
   mass(ntotal + n_bound) = rho (ntotal + n_bound) * dx * dy
   T_0(ntotal + n_bound)= TW
 END DO

!----------------Defining a line of particles on the Right side and setting
the physical properties
 DO i = 1, 2*np_side+1
   n_bound = n_bound + 1
   x(ntotal + n_bound,1) = LX
   x(ntotal + n_bound,2) = (i-1)*dx/2
   rho (ntotal + n_bound) = 1.
   mass(ntotal + n_bound) = rho (ntotal + n_bound) * dx * dy
   T_0(ntotal + n_bound)= TE
 END DO

MAX_Y= 0.
MAX_Y =0.
!------------------------ Output Files----------------------------
DO i=ntotal+1, ntotal+ n_bound

  IF (x(i,1).GT.MAX_X) MAX_X = x(i,1)
  IF (x(I,2).GT.MAX_Y) MAX_Y = x(i,2)

  WRITE (01,1016) (x(i,d), d = 1, 2)
  n_global = ntotal + n_bound
  WRITE (02,1016) (x(i-ntotal,d), d = 1, 2)
  WRITE (03,654) T_0(i)

END DO
1016 format(2(1x,D21.14))
654 format(1x,D21.14)

OPEN(17,FILE='output/GEOMETRY.DAT')
WRITE(17,*) 0., MAX_X
WRITE(17,*) 0., MAX_Y
 CLOSE(17)

END SUBROUTINE boundary_particles

!********************** NEIGHBOUR SEARCH ************************

SUBROUTINE neighbour_search(aux_p, x, n_global, ntotal, n_bound, neighbour,
dx, dy, support_radius, dist)

!---------------------------------------------------------------
! This subroutine finds the direct search for neighbour particles ! of
each fixed particle
!
! ntotal - number of particles at domain
! n_bound - number of boundary particles
! n_global - sum of the number of particles at domain and
! boundary particles
! hsml - smoothing length
! x - positions of all particles
```

```
! n_neigh - number of neighbour particles
! dist - distance between a fixed particle and a neighbour particle
! neighbour - matrix of neighbour particles
!------------------------------------------------------------------

IMPLICIT NONE

INTEGER aux_p, n_global, ntotal, n_bound, n_neigh, i, j
INTEGER neighbour(aux_p,150)
DOUBLE PRECISION r , dx_local, dy_local, support_radius
DOUBLE PRECISION x(aux_p,2), dist(aux_p,150), dx(aux_p,150), dy(aux_p,150)

DO i=1,aux_p
 DO j=1,150
   neighbour(i,j)=0
   dist(i,j) = 0.
   dx(i,j)=0.
   dy(i,j)=0.
  END DO
 END DO

 DO i=1,ntotal
    n_neigh = 1
    neighbour(i,n_neigh)=i

   DO j = 1, n_global
     dx_local = x(i,1) - x(j,1)
     dy_local = x(i,2) - x(j,2)
     r = sqrt(dx_local**2 + dy_local**2)

     IF (r.LE.support_radius) THEN

       n_neigh = n_neigh +1
       neighbour(i,n_neigh)=j
       dx(i,n_neigh) = dx_local
       dy(i,n_neigh) = dy_local
       dist(i,n_neigh) = r
     END IF

   END DO
 END DO

 OPEN(222,file='output/NEIGHBOURING_PARTICLES.dat')
 REWIND(222)

 DO i = 1, ntotal
  n_neigh = 1
   DO WHILE (n_neigh.ne.150)
     WRITE (222,150), neighbour(i,n_neigh)
     n_neigh = n_neigh +1
   END DO
 END DO

 150 format(I10)
  CLOSE(222)

 END SUBROUTINE neighbour_search
```

```
!****************** LAPLACIAN OF TEMPERATURE *********************

SUBROUTINE SPH_Temperature_Laplacian(aux_p, int_function, factor, hsml,
ntotal, dist, mass, rho, support_radius, neighbour, T_0, temp_Laplacian,
itimestep)

!------------------------------------------------------------------
! This subroutine makes calculations of the divergent of the heat
! flux !using a kernel chosen by user
!
! dist - Distance between a fixed particle and a neighbouring
! particle
! ntotal - number of particles at domain
! k_scale - scaling factor (depends on kernel)
! factor - normalization constant of kernel
! temp_Laplacian - Laplacian of temperature provided by SPH
! method
! int_function - kernel interpolation
!
! Kernel options:
! 1. Lucy's quartic kernel
! 2. Cubic spline kernel
! 3. New quartic kernel
! 4. Quintic spline kernel
!------------------------------------------------------------------

IMPLICIT NONE

INTEGER aux_p, i, j, k, d, ntotal, n_global, n_neigh, int_function,
itimestep
INTEGER neighbour(aux_p,150)
DOUBLE PRECISION temp_Laplacian (aux_p), aux, q, factor, r,
support_radius, k_scale
DOUBLE PRECISION rho(aux_p), mass(aux_p), hsml(aux_p), T_0(aux_p)
DOUBLE PRECISION dist(aux_p,150), dx(aux_p,150), dy(aux_p,150), PI

PI=ACOS(-1.)

IF (itimestep.EQ.1) THEN

  DO i=1,aux_p
   hsml(i) = 0.
  END DO

  DO i=1,ntotal

   IF (int_function.EQ.1) THEN !----------LUCY'S QUARTIC KERNEL
      k_scale = 1.0
      hsml(i) = support_radius/k_scale
      factor = 5.0/(PI*hsml(i)*hsml(i))
   ELSE
      IF (int_function.EQ.2) THEN !--------- CUBIC SPLINE KERNEL
        k_scale = 2.0
        hsml(i) = support_radius/k_scale
        factor = 15./(7.*PI*hsml(i)*hsml(i))
      ELSE
        IF (int_function.EQ.3) THEN !--------NEW QUARTIC KERNEL
```

```
          k_scale = 1.0
          hsml(i) = support_radius/k_scale
          factor = 15./ (7.*PI*hsml(i)*hsml(i))
        ELSE
          IF (int_function.EQ.4) THEN !-----QUINTIC SPLINE KERNEL
            k_scale = 3.0
            hsml(i) = support_radius/k_scale
            factor = 7./(478.*PI*hsml(i)*hsml(i))
          END IF
        END IF
      END IF
    END IF

  END DO

END IF

!---------- SPH TEMPERATURE LAPLACIAN CALCULUS(in the polar coordinates
system)

DO i=1,ntotal
  n_neigh =2
  temp_Laplacian(i)=0.

  DO WHILE (neighbour(i,n_neigh).ne.0)
    j= neighbour(i,n_neigh)
    r = dist(i,n_neigh)
    q = r/hsml(i)

!-----------------------------------------------------------------
!--------------------LUCY'S QUARTIC KERNEL-----------------------
!-----------------------------------------------------------------

    IF (int_function.EQ.1) THEN

      aux = 2*((T_0(i)-T_0(j))) * factor * ( (-12./hsml(i)**2.) +
(24.*r)/hsml(i)**3. - (12*r**2.)/hsml(i)**4. ) * (mass(j)/rho(j))
      temp_Laplacian(i)= temp_Laplacian(i) + aux
    END IF

!-----------------------------------------------------------------
!--------------------CUBIC SPLINE KERNEL-------------------------
!-----------------------------------------------------------------

    IF (int_function.eq.2) THEN

      IF ((q.ge.0.).and.(q.le.1.)) THEN

        aux = 2.*(( T_0(i)- T_0(j))) * factor * ( (-2.)/(hsml(i)**2.) +
(3.*r)/(2.*hsml(i)**3.) ) * (mass(j)/rho(j))
        temp_Laplacian(i)= temp_Laplacian(i) + aux

      ELSE
        IF ((q.gt.1.).and.(q.le.2.)) THEN
          aux = 2.*( (T_0(i)- T_0(j)) ) * factor * (-1./(2.*hsml(i))) *
((2. - (r/hsml(i) ) )**2 ) * (1./r) * (mass(j)/rho(j))
          temp_Laplacian(i)= temp_Laplacian(i) + aux
```

```
        END IF
      END IF
    END IF

!-----------------------------------------------------------------
!-------------------NEW QUARTIC KERNEL-------------------------
!-----------------------------------------------------------------
    IF (int_function.EQ.3) THEN

      aux = 2*((T_0(i)-T_0(j))) * factor * ( -18./(8.*hsml(i)**2.) +
(57.*r)/(24.*hsml(i)**3) - (20.*r*r)/(32.*hsml(i)**4.) ) * (mass(j)/rho(j))
      temp_Laplacian(i)= temp_Laplacian(i)+ aux

      END IF

!-----------------------------------------------------------------
!-------------------QUINTIC SPLINE KERNEL----------------------
!-----------------------------------------------------------------

    IF (int_function.EQ.4) THEN

      IF ((q.ge.0.).and.(q.le.1.)) THEN
        aux = 2.*( (T_0(i)-T_0(j)) ) * factor * ( -5./hsml(i))* ( 10.*
(r**3)/hsml(i)**4 - 24.*(r**2)/hsml(i)**3 + 24./hsml(i) ) *
(mass(j)/rho(j))
        temp_Laplacian(i)= temp_Laplacian(i) + aux

      ELSE
        IF ((q.gt.1.).and.(q.le.2.)) THEN
          aux = 2.*( (T_0 (i)- T_0 (j)) ) * factor * ( -5./hsml(i) )* ( -
5.* (r**3/hsml(i)**4) + 36.*(r**2/hsml(i)**3) - 90.*(r/hsml(i)**2) +
84./hsml(i) - 15./r )* (mass(j)/rho(j))
          temp_Laplacian(i)= temp_Laplacian(i) + aux

        ELSE
          IF ((q.gt.2.).and.(q.le.3.)) then
            aux = 2.*( (T_0 (i)- T_0 (j)) ) * factor * ( -5./hsml(i) )* (
(r**3)/hsml(i)**4 -(12.*r**2)/hsml(i)**3 + (54.*r)/hsml(i)**2 -
108./hsml(i) + 81./r ) * (mass(j)/rho(j))
            temp_Laplacian(i)= temp_Laplacian(i) + aux
          END IF
        END IF
      END IF
    END IF

    n_neigh =n_neigh +1

  END DO

END DO

end SUBROUTINE SPH_Temperature_Laplacian

!****************** TEMPORAL INTEGRATION ***********************

SUBROUTINE temporal_integration(aux_p, itimestep, x, ntotal, dt,
temp_Laplacian, T_0, T, step_out)
```

```
!--------------------------------------------------------------------
! This subroutine performs the integration of the temperature of ! each
particle at the domain
!
! ntotal - total number of particles at domain
! dt - timestep
! T_0 - temperature at current iteration
! T - temperature in the next iteration
!--------------------------------------------------------------------

IMPLICIT NONE

INTEGER aux_p, ntotal, itimestep, d, step_out, i, j, aux, funit
DOUBLE PRECISION temp_Laplacian(aux_p), T_0(aux_p), T(aux_p), x(aux_p,2),
dt
CHARACTER(LEN=20) aux2
CHARACTER(LEN=35) nome
OPEN(unit=66, file='output/VERIFYING_TEMP_SPH.dat')
REWIND(66)

!--------CALCULATION OF TEMPERATURE OF EACH PARTICLE AT DOMAIN---

 DO i=1,ntotal
  T(i)= T_0(i) + temp_Laplacian(i)*dt
 END DO

!------------------------- OUTPUT FILES ------------------------
aux=itimestep

IF ((aux.EQ.1).OR.(mod(aux,step_out).eq.0)) THEN
  WRITE (aux2,*) aux
  funit=aux+150

  IF (aux.LT.10)
nome='output/temperature/000000'//trim(adjustl(aux2))//'.dat'
  IF ((aux.GE.10).AND.(aux.LT.100))
nome='output/temperature/00000'//trim(adjustl(aux2))//'.dat'
  IF ((aux.GE.100).AND.(aux.LT.1000))
nome='output/temperature/0000'//trim(adjustl(aux2))//'.dat'
  IF ((aux.GE.1000).AND.(aux.LT.10000))
nome='output/temperature/000'//trim(adjustl(aux2))//'.dat'
  IF ((aux.GE.10000).AND.(aux.LE.100000))
nome='output/temperature/00'//trim(adjustl(aux2))//'.dat'
  IF ((aux.GE.100000).AND.(aux.LT.1000000))
nome='output/temperature/0'//trim(adjustl(aux2))//'.dat'
  IF (aux.EQ.1000000)
nome='output/temperature'//trim(adjustl(aux2))//'.dat'

   OPEN(funit,file=nome,status='unknown')

   DO i = 1, ntotal
     WRITE (funit,89) i, x(i,1), x(i,2), T(i)
   END DO

   CLOSE (funit)

END IF
```

```
DO i=1,ntotal
  WRITE (66,89) i, (x(i,d), d = 1, 2), T(i)
END DO

89 format(1x,I10, 1x,D21.14, 1x,D21.14, 1x,D21.14)
  CLOSE(66)

OPEN(UNIT=26,FILE='STEP_OUT.DAT')
WRITE (26,*) step_out
 CLOSE (26)

END SUBROUTINE temporal_integration

!***********************CONVERGENCE ***************************

SUBROUTINE convergence(aux_p, itimestep,T_0,T,eps,response, ntotal)

!---------------------------------------------------------------
! This subroutine verifies the convergence of the SPH solution
! Convergence criteria: abs(T_0(i) - T(i)) $<$= accuracy
! where i each particle at domain
!
! T_0 - temperature at current iteration
! T - temperature in the next iteration
! ntotal - total number of particles at domain
! dt - timestep
!---------------------------------------------------------------

IMPLICIT NONE
INTEGER aux_p, itimestep, I, ntotal
DOUBLE PRECISION T_0(aux_p), T(aux_p), error(ntotal), eps, max_error
CHARACTER(LEN=15) response

DO I=1,ntotal
  error(I) = abs(T(I) - T_0(I))

END DO

max_error = maxval(error)

print*, 'MAXIMUM ERROR = ', max_error, ' AT ', itimestep, 'a. ITERATION'

  IF (max_error.GT.eps) THEN
    response ='non_convergence'
    itimestep = itimestep+1
  ELSE
    response = 'convergence'
  END IF

!------------- UPDATING TEMPERATURES TO THE NEXT ITERATION -------
DO I=1,ntotal
  T_0(I) = T(I)
END DO

END SUBROUTINE convergence
```

```
!************ POINT-TO-POINT TEMPERATURE DIFFERENCE **************

SUBROUTINE temperature_differences(itimestep, dt, ntotal, step_out)

!-----------------------------------------------------------------
! This subroutine calculates the differences between the
! temperatures obtained by the SPH method and provided by series
!
!-----------------------------------------------------------------

INTEGER i, d, ntotal, itimestep, part_higher, part_smaller, step_out
DOUBLE PRECISION series_solution(ntotal,4), SPH_solution(ntotal,4),
Temp_differences(ntotal), smaller_difference, higher_difference, dt

OPEN(unit=11,file='output/VERIFYING_TEMP_SERIES.DAT')
OPEN(unit=66,file='output/VERIFYING_TEMP_SPH.dat')
OPEN(unit=73,file='output/TEMPERATURE_DIFFERENCES.DAT')
OPEN(unit=74,file='output/FINAL_DIFERENCES.DAT')
OPEN(unit=75,file='output/STEP_DIFFERENCES.DAT')
REWIND(73)
REWIND(74)

DO i=1,ntotal
  read(11,*) (series_solution(i,d), d=1,4)
  read(66,*) (SPH_solution(i,d), d=1,4)
END DO

DO i=1,ntotal
 Temp_differences(i)= abs(series_solution(i,4)- SPH_solution(i,4))
 WRITE (73,55) (series_solution(i,d), d=2,3), Temp_differences(i)
 WRITE (74,55) (series_solution(i,d), d=2,3), Temp_differences(i)

 IF (mod(itimestep,step_out).eq.0) THEN
   WRITE (75,55) (series_solution(i,d), d=2,3), Temp_differences(i)
 END IF

END DO
55 format (1x,D21.14, 1x,D21.14, 1x, D21.14)

higher_difference = 0.
smaller_difference = 1500.

DO i=1,ntotal
  IF (Temp_differences(i).GT.higher_difference) then
    part_higher =i
    higher_difference = Temp_differences(i)
  END IF

  IF (Temp_differences(i).LT.smaller_difference) then
    part_smaller = i
    smaller_difference = Temp_differences(i)
  END IF
END DO

WRITE (73,*) ''
WRITE (73,*) 'RESULTS AT THE ', itimestep , 'a. ITERATION'
WRITE (73,*) 'TIME = ' , itimestep*dt, ' seconds'
```

```
WRITE (73,*) ''
WRITE (73,*) '---------------------------------------'
WRITE (73,*) 'HIGHER TEMPERATURE DIFFERENCE'
WRITE (73,*) '---------------------------------------'
WRITE (73,*) ' Position = ', (series_solution(part_higher,d),d=2,3)
WRITE (73,*) ' Series Temperature --- SPH Temperature --- Difference'
WRITE (73,*) series_solution(part_higher,4), SPH_solution(part_higher,4),
higher_difference
WRITE (73,*) '-----------------------------------'
WRITE (73,*) '---------------------------------------'
WRITE (73,*) 'SMALLER TEMPERATURE DIFFERENCE'
WRITE (73,*) '---------------------------------------'
WRITE (73,*) ' Position = ', (series_solution(part_smaller,d),d=2,3)
WRITE (73,*) ' Series Temperature --- SPH Temperature --- Difference'
WRITE (73,*) series_solution(part_smaller,4),
SPH_solution(part_smaller,4), smaller_difference
WRITE (73,*) '-------------------------------------'

 CLOSE(11)
 CLOSE(66)
 CLOSE(73)
 CLOSE(74)
 CLOSE(75)

END SUBROUTINE temperature_differences
```

MATLAB SCRIPT TO PLOT GRAPHS OF TEMPERATURE

This is the MATLAB script to plot graphs of temperatures distribution

from the transient to steady-state regime.

Following the instructions to obtain the output data from the

FORTRAN program and plot figures.

```
%-----------------------------------------------------------------
% Instructions:
% 1.Within the working directory containing the FORTRAN program,
% create and open a new file in the MATLAB Editor~
% 2.Copy and paste this script and save as ''name.m''
% 3.Create a subfolder called figures within the output folder
% 4.Update the path name of the working directory in this MATLAB script
% 5.Run
% 6.Figures will be save within the figures subfolder
%-----------------------------------------------------------------

%-----------------------------------------------------------------
%---------------- PLOTTING GRAPH (SERIES SOLUTION)---------------
%-----------------------------------------------------------------
clc
clear all
close all

a= load('working_directory/output/GEOMETRY.DAT');
Xm = a(1,1);
Xmx = a(1,2);
```

```
Ym = a(2,1);
Ymx = a(2,2);

b= load('working_directory/N_PART_SERIES.DAT');
part_side = sqrt(b(1,1));
dt = b(2,1);

c= load('working_directory/NUMBER_OF_ITERATIONS.DAT');
iterations = c(1,1);

d= load('working_directory/STEP_OUT.DAT');
step_out = d(1,1);

f=load ('working_directory/output/VERIFYING_TEMP_SERIES.DAT');
fr=sortrows(f);
x_plot=fr(:,2);
y_plot=fr(:,3);
T1=fr(:,4);

x_series=reshape(x_plot,part_side,part_side);
y_series=reshape(y_plot,part_side,part_side);
T_series=reshape(T1,part_side,part_side);

figure(1);
contourf(x_series, y_series, T_series, 20);
grid on
daspect ([ 1 1 1 ]);
colormap jet;
cH = colorbar;
set(gcf, 'Position', get (0,'Screensize'));
set(cH,'FontSize',12);
set(get(cH,'title'),'string','T(^{o}C)','FontSize',12);
caxis([0 100]);
axis ([Xm Xmx Ym Ymx]);
set(gca,'xticklabel',num2str(get(gca,'ytick')','%.2f'),'fontsize',10);
set(gca,'yticklabel',num2str(get(gca,'ytick')','%.2f'),'fontsize',10);

title('SOLUTION BY SERIES','fontsize',14);
xlabel('x (m)','fontsize',12);
ylabel('y (m)','fontsize',12);
disp('Figures will be saved in working_directory/output/figures')
disp('Press any key to continue...');
pause

dir_out = (['cd working_directory/output/figures']);
eval(dir_out);

img = getframe(gcf);
imwrite(img.cdata, ['SERIES_SOLUTION.png']);
close;

%----------------------------------------------------------------
%----------------PLOTTING GRAPHS (SPH SOLUTIONS)------------------
%----------------------------------------------------------------
dir_in = (['cd working_directory/output/temperature']);
%----------------------INITIAL DISPOSITION------------------------
```

```
FileName=(['0000001']);
t = load ('working_directory/output/temperature/0000001.dat');

fr=sortrows(t);
x_plot=fr(:,2);
y_plot=fr(:,3);
T1=fr(:,4);

x_SPH =reshape(x_plot,part_side,part_side);
y_SPH =reshape(y_plot,part_side,part_side);
T_SPH=reshape(T1,part_side,part_side);

eval(dir_out);
figure(2);
contourf(x_SPH, y_SPH, T_SPH, 20);

grid on
daspect ([ 1 1 1 ]);
colormap jet;
cH = colorbar;
set(gcf, 'Position', get (0,'Screensize'));
set(cH,'FontSize',12);
set(get(cH,'title'),'string','T(^{o}C)','FontSize',12);
caxis([0 100]);
axis ([Xm Xmx Ym Ymx]);
set(gca,'xticklabel',num2str(get(gca,'ytick')','%.2f'),'fontsize',10);
set(gca,'yticklabel',num2str(get(gca,'ytick')','%.2f'),'fontsize',10);

title1 = (['SPH SOLUTION - Cubic Spline Kernel - t = 0.00 s']);
title(title1,'Units','Normalized','fontsize',12)
title(title1,'fontsize',12,'fontweight','b')
xlabel('x (m)','fontsize',12);
ylabel('y (m)','fontsize',12);

img = getframe(gcf);
imwrite(img.cdata, [FileName, '.png']);
clear FileName;

for i = step_out:step_out:iterations,
    num=int2str(i);
    char1=num2str(i);
    char2=num2str(iterations);
    eval(dir_in);

      if (i $>$= 0) & (i $<$ 10)
           FileName=([ '000000' num]);
        else
          if (i $>$= 10) & (i $<$ 100)
            FileName=([ '00000' num]);
          else
            if (i $>$= 100) & (i $<$ 1000)
              FileName=([ '0000' num]);
            else
              if (i $>$= 1000) & (i $<$ 10000)
                FileName=([ '000' num]);
               else
                 if (i $>$= 10000) & (i $<$ 100000)
```

```
              FileName=([ '00' num]);
              else
                if (i $>$= 100000) & (i $<$ 1000000)
                FileName=(['0' num]);
                else
                   FileName=([ num]);
                 end
               end
             end
           end
         end
      end

       disp(['opening' FileName '.dat' ]);
       t = load ([FileName '.dat']);

       fr=sortrows(t);
       x_plot=fr(:,2);
       y_plot=fr(:,3);
       T1=fr(:,4);

       clear FileName;

       x_SPH =reshape(x_plot,part_side,part_side);
       y_SPH =reshape(y_plot,part_side,part_side);
       T_SPH=reshape(T1,part_side,part_side);

       figure(i);
       contourf(x_SPH, y_SPH, T_SPH, 20);

       grid on
       daspect ([ 1 1 1 ]);
       colormap jet;
       cH = colorbar;
       set(gcf, 'Position', get (0,'Screensize'));
       set(cH,'FontSize',12);
       set(get(cH,'title'),'string','T(^{o}C)','FontSize',12);
       caxis([0 100]);
       axis ([Xm Xmx Ym Ymx]);
set(gca,'xticklabel',num2str(get(gca,'xtick')','%.2f'),'fontsize',10);
set(gca,'yticklabel',num2str(get(gca,'ytick')','%.2f'),'fontsize',10);

       aux2=i*dt;
       num1=num2str(aux2);
       title1 = (['SPH SOLUTION - t = ' num1 ' s']);
       title(title1,'Units','Normalized','fontsize',12)
       title(title1,'fontsize',12,'fontweight','b')
       xlabel('x (m)','fontsize',12);
       ylabel('y (m)','fontsize',12);

       eval(dir_out);
       img = getframe(gcf);
       if i $<$ 10
         nome=(['000000' num '.png']);
         saveas(gcf,nome);
       end
       if (i $>$= 10) & (i $<$ 100)
```

```
        nome=(['00000' num '.png' ]);
        imwrite(img.cdata, nome);
      end
      if (i $>$= 100) & (i $<$ 1000)
        nome=(['0000' num '.png' ]);
        imwrite(img.cdata, nome);
      end
      if (i $>$= 1000) & (i $<$ 10000)
        nome=(['000' num '.png' ]);
        imwrite(img.cdata, nome);
      end
      if (i $>$= 10000) & (i $<$ 100000)
        nome=(['00' num '.png' ]);
        imwrite(img.cdata, nome);
      end
      if (i $>$= 100000) & (i $<$ 1000000)
        nome=(['0' num '.png' ]);
        imwrite(img.cdata, nome);
      end
      if (i $>$= 1000000)
        nome=([ num '.png']);
        imwrite(img.cdata, nome);
      end
   close all
end
```

References

1. Fraga Filho, C.A.D.: Development of a computational instrument using a Lagrangian particle method for physics teaching in the areas of fluid dynamics and transport phenomena. Rev. Bras. Ensino Fís. (online) **39**(4), e4401 (2017). https://doi.org/10.1590/1806-9126-rbef-2016-0289
2. Coveney, P.V., Boon, J.P., Succi, S.: Bridging the gaps at the physics-chemistry-biology interface. Philos. Trans. R. Soc. A (2016). https://doi.org/10.1098/rsta.2016.0335
3. Stalter, S., Yelash, L., Emamy, N., Statt, A., Hanke, M., Lukácové-Medvid'ová, M., Virnau, P.: Molecular dynamics simulations in hybrid particle-continuum schemes: pitfalls and caveats. Comput. Phys. Commun. **224**, 198–208 (2018)
4. Liu, W.K., Park, H.S., Qian, D., Karpov, E.G., Hiroshi, K., Wagner, G.J.: Bridging scale methods for nanomechanics and materials. Comput. Methods Appl. Mech. Eng. **195**, 1407–1421 (2006)
5. Petsev, N.D., Leal, L.G., Shell, M.S.: Multiscale simulation of ideal mixtures using smoothed dissipative particle dynamics. J. Chem. Phys. **144**, 084115 (2016). https://doi.org/10.1063/1.4942499
6. Liu, M.B., Liu, G.R., Zhou, L.W., Chang, J.Z.: Dissipative particle dynamics (DPD): an overview and recent developments. Arch. Comput. Methods Eng. **22**, 529–556 (2015)
7. Hemminger, J.: From quanta to the continuum: opportunities for mesoscale science. A Report from the Basic Energy Sciences Advisory Subcommittee, U.S. Department of Energy, USA (2012). https://doi.org/10.2172/1183982
8. Delgado-Buscalioni, R., Coveney, P.V., Riley, G.D., Ford, R.W.: Hybrid molecular-continuum fluid models: implementation within a general coupling framework. Philos. Trans. R. Soc. A **363**(1833), 1975–85 (2005)
9. Borgh, M.K., Lockerby, D.A., Reese, J.M.: Fluid simulations with atomistic resolution: a hybrid multiscale method with field-wise coupling. J. Comput. Phys. **255**, 149–165 (2013)
10. Sih, G.C. (ed.): Multiscaling in Molecular and Continuum Mechanics: Interaction of Time and Size from Macro to Nano: Application to Biology, Physics, Material Science, Mechanics, Structural and Processing Engineering. Springer, The Netherlands (2007)
11. Mukhopadhyay, S., Abraham, J.: A particle-based multiscale model for submicron fluid flows. Phys. Fluids **21**, 027102 (2009). https://doi.org/10.1063/1.3073041
12. Yamaguchi, T., Ishikawa, T., Imai, Y., Matsuki, N., Xenos, M., Deng, Y., Bluestein, D.: Particle-based methods for multiscale modeling of blood flow in the circulation and in devices: challenges and future directionsD. Ann. Biomed. Eng. **38**(3), 1225–35 (2010)
13. Ren, W., Weinan, E.: Heterogeneous multiscale method for the modeling of complex fluids and micro-fluidics. J. Comput. Phys. **204**, 1–26 (2005)

C. A. Dutra Fraga Filho, *Smoothed Particle Hydrodynamics*,
https://doi.org/10.1007/978-3-030-00773-7

14. Chen, S., Wang, W., Xia, Z.: Multiscale fluid mechanics and modeling. In: Bai, Y., Wang, J., Fang, D. (eds.) Mechanics for the World: Proceedings of the 23rd International Congress of Theoretical and Applied Mechanics, ICTAM2012, Beijing, China, Procedia IUTAM, vol. 10, pp. 110–114 (2014)
15. Moeendarbary, E., Ng, T.Y., Zangeneh, M.: Dissipative particle dynamics: introduction, methodology and complex fluid applications—a review. Int. J. Appl. Mech. **1**(4), 737–763 (2009)

Index

B
Boundary treatment techniques, 53
  artificial repulsive forces, 54
  dynamic particles, 56
  implementations, 67
  Lennard-Jones molecular force, 54
  open periodic boundary conditions, 60
  reflective boundary conditions, 56
  virtual particles, 54

C
Classical laws of physics, 1
Consistency, 46
  correction of the pressure gradient, 50
  CSPM method, 48
  density renormalization, 50
  free surface, 48
  kernel gradient correction method, 52
  mean least squares (MLS), 51
  particle inconsistency, 47
  restoration of the consistency, 48
Continuum hypothesis
  continuum mechanics, 11
  laws of physics, 11
  molecular motion, 12
Continuum mechanics, 1, 60

D
Dam breaking
  artificial viscosity, 86
  coefficient of restitution of kinetic energy, 87
  CSPM method, 86
  cubic spline kernel, 86
  density correction, 86
  dynamic pressure, 86, 92
  experimental results, 88
  Fay's adjusted curves, 93
  free surface, 86
  kernel gradient correction, 86
  oil slick diameter, 93
  pressure of the oil column, 92
  pressure of the water column, 86
  pressure on the free surface, 86
  reflective boundary conditions, 88
  wave front, 87
  XSPH correction, 86
Dirac delta function, 18
Discretization of the domain
  continuum mechanics, 19
  2-D cell, 20
  laws of physics, 19
  microscopic molecular fluctuations, 19
  molecular fluctuations, 19
  radius of the domain of influence, 20
Dissipative Particle Dynamics (DPD), 105
Divergent of a vectorial function, 20
  integration surface, 21

E
Eulerian view, 1
  complex geometries, 2
  free surface, 2
  remeshing, 2

C. A. Dutra Fraga Filho, *Smoothed Particle Hydrodynamics*,
https://doi.org/10.1007/978-3-030-00773-7

topological changes, 2
External forces, 13
free surface forces, 101
gravitational force, 13
Lennard-Jones molecular force, 101

F

Fluid at rest inside the reservoir
analytical solution, 83
collision detection algorithm, 80
cubic spline kernel, 81
hydrostatic equilibrium, 82
immobile reservoir, 78
isothermal liquid, 80
modified pressure, 80
modified pressure gradient, 80
quintic spline kernel, 81
SPH errors, 79
Free surface, 29, 34, 48, 50, 53, 54, 79, 82, 86, 101
Fundamentals of the SPH
Dirac delta function, 18
domain of influence, 18

G

General SPH computer code, 101
Gradient of a scalar function, 22, 23
Gradient of the kernel
properties
anti-symmetry, 116
non-normalization, 116

H

Homogeneous flat plate
artificial repulsive forces, 71
cubic spline kernel, 72
Dirichlet boundary conditions, 69
heat diffusion, 68
kernel truncation, 76
Lucy's quartic kernel, 72
new quartic kernel, 72
particle inconsistency, 76
quintic spline kernel, 72
solution by series, 70
steady-state, 69
temperature distribution, 70
Hybrid molecular-continuum model, 105

L

Lagrangian modelling, 2
conservation of mass, 2
lagrangian element, 2
meshless, 2
numerical diffusion, 2
particles' history, 2
processing time, 2
Laplacian
deduction of the SPH Laplacian, 115
Laplacian of a function, 24
Laplacian of temperature, 69

M

Macroscopic domain, 1
Molecular dynamics, 1
Multiscale model, 105

N

Neighbouring particles search, 28
cell linked list, 28
direct searching, 28
grid, 28
Numerical aspects
artificial viscosity, 40
Numerical aspects and corrections
artificial viscosity, 40
energy conservation equation, 41
momentum conservation equation, 41
particles aglomerates, 42
tensile instability, 42
XSPH correction, 42

O

Oil spreading on a calm sea
artificial viscosity, 95
collisions detection and response, 96
CSPM method, 95
cubic spline kernel, 95
dam break, 91
density renormalization, 95
Fay's equation, 90
gravity-inertial regime, 90
interface plane, 92
non-penetration condition, 92
oil slick reaching, 90
prediction model, 90
rough and rigid plane, 92
smoothing length, 95
turbulence, 96
XSPH correction, 96

P

Particle inconsistency, 47, 76

Physical laws of conservation, 12
  Lagrangian view, 12
  Law of conservation of mass, 13
  law of energy conservation, 13
  law of momentum conservation, 13
Pressure modelling, 13
  dynamic pressure, 14
  equation of state, 13
  free surface flows, 13
  free surfaces, 15
  hydrostatic pressure, 14
  mach number, 14
  modified pressure, 14
  tait equation, 13

R

Reflective boundary conditions, 56
  coefficient of friction, 57
  coefficient of restitution of kinetic energy, 57
  collision detection and response, 57
  elasticity, 57
  friction, 58
  Newton's law of restitution, 80
  non-penetration condition, 56

S

Scientific method, 3
Smoothing functions, 26
  cubic spline kernel, 27, 108
  derivatives of the kernel, 118
  Lucy's quartic kernel, 27, 107
  new quartic kernel, 27, 110
  normalization constant, 28
  properties, 19
    compact support, 19
    convergence, 19
    decay, 19
    positivity, 19
    simmetry, 19
    smoothness, 19
    unity property, 19
  quintic spline kernel, 27, 112
  scaling factor, 19
  unity property, 114
Smoothing length, 19, 39, 95
Specific internal energy, 15
  incompressible fluid, 15
  temperature, 15
SPH approximations
  conservation equations, 25
    energy, 26
    momentum, 25, 26
SPH errors, 26
SPH fundamentals, 17
Support radius, 19, 35, 47, 101

T

Taylor series, 26, 46, 115
  remainder of Taylor series, 115
  residue of second order, 26
  truncation, 26
Temporal integration methods
  CFL numerical stability criterion, 44
  Euler's method, 44
  leap frog method, 45
  predictor-corrector, 45
Treatment of the free surface, 29
  color field, 32
  continuum surface force method, 30
  curvature, 32
  curvature of the free surface, 30
  divergence of the particle's position, 34
  free surface parameter, 34
  gradient of the smoothed color field, 32
  I-SPH method, 33
  smoothed color field, 32
  surface particle tracking, 34
  surface tension, 30
  surface tension coefficient, 30
  theshold parameter, 33
Treatment of the interfaces
  contact algorithm, 35
  intermolecular forces, 37
  Lennard-Jones molecular force, 36
  particle-particle contact, 35
  repulsive forces, 37
  restorative contact force, 36
  rigid sphere, 35
  spherical shaped particle, 35
  SPH particle, 35
  support radius, 35
Turbulence, 37
  direct numerical simulation (DNS), 37
  explicit algebraic reynolds stress model, 38
  $\kappa - \varepsilon$ model, 38
  Lagrangian averaged Navier–Stokes (LANS), 38
  large eddy simulation (LES), 37
  SPH-ε model, 38

V

Variable smoothing length, 39
  time derivative of the support radius, 40

MIX
Papier aus verantwortungsvollen Quellen
Paper from responsible sources
FSC® C105338

If you have any concerns about our products,
you can contact us on
ProductSafety@springernature.com

In case Publisher is established outside the EU,
the EU authorized representative is:
Springer Nature Customer Service Center GmbH
Europaplatz 3, 69115 Heidelberg, Germany

Printed by Libri Plureos GmbH
in Hamburg, Germany